THE NAUTICAL INSTITUTE

CRACKING THE CODE
The relevance of the ISM Code
and its impact on shipping practices

by
Doctor Philip Anderson DProf BA(Hons) FNI

with supporting chapters by:
Captain Stuart Nicholls MNI
Captain John Wright FNI
Captain Sean Noonan FNI

dedicated to
Mr. W.A O'Neil
Secretary General, International Maritime Organization

CRACKING THE CODE

by Doctor Philip Anderson DProf BA(Hons) FNI
with supporting chapters by
Captain Stuart Nicholls MNI
Captain John Wright FNI
Captain Sean Noonan FNI

Published by The Nautical Institute
202 Lambeth Road, London SE1 7LQ, England
Telephone: +44 (0)207 928 1351   Fax: +44 (0)207 401 2817
website: www.nautinst.org

First edition published 2003
Copyright The Nautical Institute 2003

The book has been prepared to address the subject of the use and implementation of the ISM Code. This should not, however, be taken to mean that this document deals comprehensively with all of the concerns which will need to be addressed, or even, where a particular matter is addressed, that this document sets out the only definitive view for all situations. The opinions expressed are those of the authors only and are not necessarily to be taken as the policies or views of any organisation with which they have any connection.

Readers should make themselves aware of any local, national or international changes to bylaws, legislation, statutory and administrative requirements that have been introduced which might affect any decisions taken on board.

Typeset by J A Hepworth
1 Ropers Court, Lavenham, Suffolk CO10 9PU, England

Printed in England by O'Sullivan Printing Corporation
Trident Way, International Trading Estate, Brent Road
Southall, Middlesex UB2 5LF

ISBN 1 8700 77 63 6

CONTENTS

FOREWORD

by Mr. W.A. O'Neil
Secretary General The International Maritime Organization

The introduction of the International Safety Management (ISM) Code was a turning point in the way that maritime administrations saw the future improvement of standards. Its origins lay in a number of high profile accidents where the ships were technically meeting all the convention and classification standards; the officers were well qualified and the crew sufficient and well trained for their roles. Sadly, despite this, systems failed with tragic loss of life or major pollution.

In one particular case — the loss of *The Herald of Free Enterprise* — because it was so thoroughly investigated, the outcome challenged the existing order to come up with a better solution. Shipping, however, was not the only industry to be having accidents which were totally unexpected. The explosion and fire on the oil platform *Piper Alpha* showed all the same symptoms which had been exhibited in the nuclear near-catastrophe at Three Mile Island. And the aviation industry has had to learn the same lessons and has procedures and practices for training pilots and maintaining airframes which are based upon human element principles. Similarly, the chemical industry, as exemplified by Du Pont, which started as a manufacturer of explosives, has incorporated risk-based safety management into its operations, which is at the heart of health and safety legislation.

The evidence in all these incidents and practices is clear. Safety cannot be left to chance. It has to be managed. Normally in life we may choose to take risks with our own safety, when crossing the road for example, but on a ship and in an international industry which is offering a public service, the exercise of this personal freedom is not acceptable if it could lead to the loss of lives or the pollution of a coastline.

It is for these reasons that I am committed to the effective implementation of the ISM Code. In my view it is one of the most significant steps forward that IMO has taken in maritime safety since its foundation, not because it supersedes that status of all the other IMO Conventions, but because it embraces their standards and provides the framework through which they can be implemented.

ii

Dr. Philip Anderson is to be congratulated on his research into the response of both seafarers and shore staff to the implementation of the Code. He has demonstrated that it has received a mixed reaction, and perhaps that should have been expected. After all, the shipping industry has good and proud traditions, not dissimilar to the aims of the Code, but the Code requires some fresh thinking and individuals who may see themselves as 'free spirits' may perhaps feel inhibited by the new strictures.

I understand this, and so too do the other authors in this well-presented book, who are able to demonstrate how safety management can become the focus of efficient operations and well-motivated crews. This is surely what we are all aiming for.

I am touched that this book has been dedicated to me. It breaks new ground in the way that it takes on a leadership role, which will inspire those who read it, and it gives me particular satisfaction, as I believe strongly that the ISM Code and the safety culture it engenders is the right way to go.

PREFACE

In the run up to the deadline for compliance for Phase One ISM ships, I was invited by Lloyds of London Press to write a practical guide to the legal and insurance implications of the ISM Code. The resulting volume, *ISM Code – A Practical Guide to the Legal and Insurance Implications* (LLP1998, ISBN 1-85978-621-9), was eventually published in December 1998. It was with a sense of some frustration that the final manuscript contained a number of important questions concerning the ISM Code which remained unanswered. However, I anticipated that within a relatively short period of time answers would be forthcoming from the courts. The sort of issues involved the implications of the role of the Designated Person, the way in which evidence produced within a working SMS might be used in legal actions and other related topics. As the deadline for Phase Two compliance started to approach the answers were still elusive and the whole question of the success or otherwise of ISM compliance was far from clear.

In an attempt to try and find some answers I took it upon myself to conduct a major piece of research across the international shipping and related industries. To introduce a level of critical objectivity and scrutiny into the research I registered the programme with the National Centre for Work Based Learning Partnerships with the Middlesex University. The findings and conclusions were to be submitted in support of an external, part-time doctorate. The research started in April 2001 with the launch of a dedicated website – www.ismcode.net and the distribution of nearly 70,000 questionnaires to seafarers, ship operators and a whole range of other related individuals and organisations around the world. Other related research projects were launched simultaneously. The aim was to try and establish how the ISM Code implementation was progressing and what issues might have come to light during the implementation process.

By November 2001 almost 3000 completed questionnaires had been returned, along with about 800 detailed narrative comments from individuals, providing an insight into their own experiences of ISM implementation. This enormous amount of data provided an extremely valuable insight into the spectrum of experiences. It revealed a complicated and complex picture. There were clearly many problems being experienced, some very serious, with implementation but there was also evidence of some very positive experiences. A whole range of misunderstandings came to light about some quite fundamental ISM issues – particularly amongst seafarers. Upon investigation to try and establish why there should be such widespread misunderstanding about very similar issues from individuals in different parts of the world, it came to light that there was very little by way of good quality information easily and readily available to seafarers to introduce them to the actual underlying principles and philosophies of the ISM Code. Somehow misconceptions had permeated and spread throughout the industry. The sort of misguided ideas which seemed to be widespread were that the ISM Code requires large quantities of paperwork and administration to function, that ticking boxes and checklists would replace good seamanship and proper training. It became apparent that part of the reason, at least, for such serious misunderstandings was that little if any instruction, guidance or training had been provided to seafarers on ISM. The first real encounter they usually had with a Safety Management System (SMS) was when a rather large set of procedure manuals arrived on board. These tended to be accompanied by a message along the lines of 'here is your SMS, now get on with it!' The SMS was completely alien to those who were now being charged with its implementation. It is little wonder therefore that there was so much resentment and negativism being expressed.

As a consequence of this extremely worrying feedback and bearing in mind that we were on the final straight in the run up to Phase Two implementation I decided that priority must be given to providing a practical set of guidelines which would introduce seafarers, and others, to the basic concepts of ISM and to try and dispel some of the myths which had evolved. Accordingly, in early 2002 I broke off from the main research programme and recruited a colleague, Peter Kidman of Intercargo, to assist in writing a suitable set of guidelines. The result was two volumes: a small book published by the North of England P&I Association – *A Seafarers Guide to ISM* (North of England P&I, 2002, ISBN 0-9542012-2-1) and a comic strip book – *What have the World Cup and ISM got in common?* (North of England P&I, 2002, ISBN 0-9542012-1-3).

Returning to the research in the Spring of 2002, considerable time was spent analysing and trying to interpret the data and other feedback. My findings and conclusions were written up and submitted to Middlesex University in November 2002. Whilst these findings and conclusions were sufficient to satisfy the requirements of the University to demonstrate an ability to conduct academic research at a high level and to push the boundaries of existing knowledge forward — sufficient to be awarded a Doctorate in Professional Studies — big questions still remained. The intention had been to publish those findings and conclusions in a style which would be non-academic but rather informative and relevant to the interested reader in the industry both at sea and ashore.

However, neither myself nor the publishers, The Nautical Institute, felt entirely comfortable with the original manuscript setting out those findings and conclusions. There was certainly much interesting information and, for the first time, empirical evidence about ISM implementation and the many problems which were being encountered. It represented a snap-shot of the status of ISM at a point in time between Phase One and Phase Two implementation. It identified many problems which were being encountered by many individuals and organisations which were inhibiting the successful implementation but provided little by way of practical advice on what could be done to overcome some of these problems. Again I had ended up with a manuscript with many unanswered problems. The industry would not only need to know what was going wrong but also be given some hope that ISM could work and to provide some suggestions on what could be done to put things right. I was determined, on this occasion, to try and close the loop and, hopefully, help to show the way forward in the book which was to be published.

From the responses to the research questionnaires and other contacts I had identified a relatively small number of individuals and organisations who had clearly had a very positive and fruitful experience of ISM. The message coming from this group was that the ISM Code had not only made their ships and companies safer, but also more efficient and, interestingly, more profitable! Clearly, if this could be demonstrated, it would provide a most powerful and encouraging message, should one be needed, to ship operators and their accountants, that the ISM Code really is worth making the proper investment in — that safe ships are profitable ships.

Accordingly, I reopened my research in the early part of 2003 to try and establish in more detail what it was that those who had had such a positive experience of ISM were doing that many of their colleagues were perhaps not doing. As a result I identified three individuals, from quite different sectors of the industry, who had had enlightening experiences and who had very important and relevant stories to tell.

Captain Stuart Nicholls had been in command at sea with a first class liner operator, where he had helped introduce their SMS and believed that, like many seafarers, they managed safety perfectly well before ISM and it really added little benefit to their operation. However. he then underwent a career change and went to work onboard an offshore mobile drilling unit. There he encountered a level of safety management far beyond anything he had ever imagined — where safety really was raised to the very highest priority beyond everything else. He does not advocate that the way safety is managed in the offshore industry should be imposed lock-stock-and-barrel on the merchant marine, but he does suggest that there are many lessons we can learn from that industry.

Captain John Wright has vast experience of shipping as well as offshore and other industries and provides management skills training to senior officers in the merchant marine — in particular Crew Resource Management (CRM) as well as Risk Management. He has almost daily encounters with a wide range of seagoing masters and officers with whom he explores management techniques which might currently exist on board many ships and he teaches the types of management technique which are anticipated by management systems such as the ISM Code.

Captain Sean Noonan was a D.P.A in a relatively small shipping company who had set up one of the most imaginative and successful SMSs I had encountered throughout my research. As a direct consequence of their ISM implementation the company saved enormous amounts of money. They devised an SMS which was plugging the many holes which existed in their organisation through which losses were pouring out.

Since each of these three individuals had their own fascinating story to tell, from first hand direct experience, I felt that they should be given the opportunity to tell their own story in their own words rather than for me to provide a second hand account. I am most grateful to each of them that they very kindly agreed to contribute their own story by way of additional chapters to follow my own findings and conclusions in this book.

A fundamental principle of the ISM Code is that each ship operator is individual and that their SMS should be developed to fit their individual organisation. However, this book may help to show how and why similar problems can arise in very different companies but will also show, hopefully, how many of those problems can be overcome.

The important lesson to learn is that ISM can work. When an SMS has been properly developed and implemented it is the most profitable business management tool which anyone could ever imagine. As time goes on this lesson is being learnt by more and more ship operators who are making the necessary investment to have an SMS that is working as it was anticipated by the original draughtsmen of the Code. However, it is clear that many other ship operators still need to learn this important lesson and be convinced that it really is worth making that commitment and investment.

I hope that all who read this book will derive some value and added knowledge and that it will help to stimulate much more debate. In particular, those companies who are experiencing good positive results from their own ISM implementation I would ask to please share that good experience with their colleagues in the industry — for the advancement of safer ships, cleaner seas and increased efficiency, profits and prosperity for all who rely on shipping for their daily bread!

Phil Anderson, Acomb, July 2003

ACKNOWLEDGEMENTS

One of my biggest regrets of the entire research programme, which has culminated in this book, is that I could not thank every single contributor personally for their participation and for sharing their experiences with me. I received nearly 3,000 completed questionnaires – both on-line and paper copies and about 800 detailed narrative responses – ranging from a few lines to very detailed accounts running to thousands of words. I received ISM jokes, horror stories and even an 'ISM fairy story'! Many of the respondents clearly had very bad experiences of ISM, for which I felt sadness but live in hope that they will have an opportunity to experience ISM as it was intended to work and realise the benefits. To all those who responded and shared their experiences I thank you all and apologise that I cannot thank you individually and personally.

There have been so many individuals and organisations who have helped me during the research that, given the best will in the world, I cannot list them all and give the due credit they all justly deserve. My gratitude to all those who helped is extended in a most sincere way and if I have omitted to mention them individually here then it is purely on account of keeping this acknowledgement within an acceptable size. Having said that there are a number of individuals and organisations who I feel I must single out – but not in any particular order:

- The Nautical Institute, as publishers, who assisted with some practical help which allowed me to partially fund an enormous printing and postage bill related to the mailing out of the questionnaires. In particular, Julian Parker for all his help and support throughout — especially with the pulling together of the final manuscript. Also special mention to Claire Walsh and the Institute journal *SEAWAYS* for support in promoting the project.

- The North East branch of the Nautical Institute for their support in sponsoring the dedicated website www.ismcode.net

- Various Nautical Institute branches around the world who conducted local debates and research into ISM experiences, with feedback which allowed an interesting understanding of regional perspectives.

- The Marine Society for their practical help and support, which also helped towards the funding of the questionnaires and mailshot.

- The North of England P&I Association who very kindly covered the University fees.

- Dr. Kathy Doncaster and Professor John Stephenson of Middlesex University for their guidance and supervision throughout the academic project, as well as Mr Chris Hilton as external supervisor for support and keeping me on track.

- Rob Errington of Remedia for assistance above and beyond the call of duty in setting up the website and integrated relational database — as well as exercising such considerable patience teaching me how to use and extract information from Access!

- Yuquing Ye for her help, patience and perseverance in manually inputting the data from about 2000 paper questionnaires into the Access database.

- Members of staff at North of England P&I association who assisted in many ways but in particular with some enormous mailshots. In particular Adele and Denise and Marcia and her excellent team.

- *Lloyds List* for placing a banner on the front page of their own webpage encouraging their readers to participate in the survey as well as their support throughout with many helpful editorial articles.

- NUMAST and *The Telegraph* for support and assistance in distributing copies of the questionnaire to their members as well as extensive editorial coverage.

- IMerEST and MER for editorial coverage and encouraging their members to participate in the survey.

- The Mission to Seafarers, *The Sea* and in particular Gillian Ennis and the Rev. Cannon Ken Peters for considerable encouragement and support and for distributing very significant numbers of the questionnaire to missions around the world and to the local chaplains who took the questionnaires on board the ships they visited. This intervention meant that many additional seafarers of different nationalities and cultures were able to share their experiences of ISM implementation.

- The many other shipping related newspapers, magazines and journals who carried editorial coverage and helped open up the international debate on ISM implementation.

- The IMO and the Secretary General — Mr William O'Neil — for having the vision and commitment to introduce the ISM Code, support me generally during the research programme and for the foreword to this book. Also a special thanks to Captain Andy Wimbow for much practical assistance and guidance.

- BIMCO for actively promoting the project and for encouraging their own members to participate — through links on their website as well as through their publications.

- Captain Stuart Nicholls, Captain John Wright and Captain Sean Noonan, for their valuable advice and guidance and especially for their respective chapters at the end of this book.

- My wife Val and sons Chris, Matty and Stu for putting up with me during these last few years whilst holding down a very demanding full time job, a part time job as well as undertaking the research project, preparing the doctoral submission and writing this book. Hopefully we can now spend a little more time together!

Dedication to IMO Secretary-Genereal Mr. William A. O'Neil

The year of publication of this book coincides with the retirement of one of the truly great 'safety-men' of shipping of this era - Mr William O'Neil, Secretary - General of IMO. Since 1990 Bill O'Neil has pioneered many important safety and pollution related initiatives through the various developmental stages of IMO. However, I believe it will be his work in bringing the human element into focus - as the major factor in accidents and losses - which will be his true legacy. Of particular relevance are the SCTW Convention and, within the context of this book, the ISM Code. There may still be some way to go - but we are heading in the right direction.

On behalf or myself, my co-authors and The Nautical Institute I would like to dedicate this book to William O'Neil in gratitude for his foresight and hard work during his period of office which has helped move our industry forward towards making our ships safer and seas cleaner.

Phil Anderson - Acomb, Northumberland September 2003.

THE OBJECTIVES OF THE INTERNATIONAL SAFETY MANAGEMENT (ISM) CODE

PREAMBLE

"The purpose of this Code is to provide an international standard for the safe management and operation of ships and for pollution prevention".

3.2 OBJECTIVES

3.2.1 The objectives of the Code are to ensure safety at sea, prevention of human injury or loss of life, and avoidance of damage to the environment, in particular to the marine environment and to property.

3.2.2 Safety-management objectives of the Company should, *inter alia*:

  .1   Provide for safe practices in ship operation and a safe working environment.

  .2   Establish safeguards against all identified risks.

  .3   Continuously improve safety-management skills of personnel ashore and board ships, including preparing for emergencies related both to safety and environmental protection.

3.2.3 The safety-management system should ensure:

  .1   Compliance with mandatory rules and regulations.

  .2   That applicable Codes, guidelines and standards recommended by the organisation, administrations, classification societies and maritime industry organisations are taken into account.

Chapter 1

THE RELEVANCE OF THE ISM CODE AND ITS IMPACT ON SHIPPING PRACTICES

Introduction

The primary role of shipping is to make a profit from carrying and delivering cargo for freight. So, when the industry is compelled to introduce new safety management systems into operational practices, there could easily be some misunderstanding about the need.

Seafarers, after all, know that dangers should be avoided and that the Company will show little sympathy in the event of a collision or stranding. Customs of the trade, however, tend to cover losses and damage, which are taken care of through insurance. The commercial priority is to minimise the liabilities to which the ship owner is exposed in the contracts that govern employment, cargo carriage, insurance and port operations. By astutely avoiding liabilities, the ship owner can avoid unnecessary losses.

There is, of course, another dimension to seafaring and that is compliance with statutory regulations. Often, these are detailed and invasive. They tell the master how to navigate, what performance standards shall apply to marine equipment, how to conduct drills and inspect the ship. Nothing, it appears, can be left to chance. Even the regulations for preventing collisions at sea define how the rules are to be interpreted and when there is an incident they are so written that fault will be assigned in proportion to blame.

Today on top of the existing regulations we now have a new safety management code instructing the industry that the most important matter to be addressed, both ashore and afloat, is that safety must come first. This is not an argument about commercial interest, but a matter of personal conviction. If we examine the statistics for loss of life and personal injuries we find that merchant shipping is the second most dangerous occupation in the world. Within these statistics we find certain companies with a very good record. Chevron, for example, can demonstrate an exemplary safety performance of less than one lost time accident per 200,000 man-hours against an industry norm estimated to be in excess of two. This means that other companies must be very much worse.

1

Avoiding ship losses and cargo damage is analogous to avoiding accident and injury to people. Work is considered, risks are assessed and sensible precautions are taken. If there are problems these can be looked at so that changes can be made to the way work is done to avoid problems in the future.

Here, then, lies the dissonance between those who conduct their affairs by complying with regulations and protecting the provisions in commercial contracts with those who define risks within the context of their working practices and use their initiative (within an experienced seaman-like framework) to avoid hazards and potentially dangerous practices.

At the time of writing there is some turmoil in the minds of those working in the shipping industry because they have been given a regulation in the International Management Code for the Safe Operation of Ships and for Pollution Prevention (International Safety Management (ISM) Code) which requires them to manage safety as an integral part of their working practices, but there are no prescriptive rules to say how this must be done, only broad guidelines.

If we look at the costs implicit in this legislation they have, in general, gone up. Companies must have a dedicated safety manager, ships must have detailed procedures and an expensive audit process is required. The mandatory nature of this regulation means that the ISM Code now applies to all companies. As a result of this, the costs and benefits have to be reassessed in terms of operational efficiency.

Bad implementation of the ISM Code produces nothing but resentment because the procedures are not relevant, working practices are costly to change and check lists burden the system to satisfy the auditing requirement. Scarce energy, time and resources are diverted from the principal purpose of carrying cargo profitably.

Good ISM Code implementation, on the other hand, is effective in reducing accidents and damage. When it is integrated into the working of the ship it provides positive encouragement to meet commercial objectives safely. A good ISM system is now undoubtedly a competitive advantage.

One of the first lessons in business management is that meeting company objectives well and efficiently must be rewarded. It is therefore difficult to understand how wrongly some companies have approached the introduction of the ISM Code. It has clearly been seen as an administrative burden so that success is defined as being able to comply with the audit. In this process the reward for good safety management is turned into a penalty because non-conformity found in the system appears

2

as a black mark against those running it. This reinforces the blame culture and effectively penalises initiative and innovative safety improvement.

There is another aspect of safety management which is not implicit in the systems themselves but which is, nonetheless, important. Society has perceptions about shipping, which are largely influenced by the media, but they also have expectations. These are usually based on their experience driving cars, flying and working in new buildings. Generally, safety performance in all these domains is improving. If there is a serious accident in shipping it creates strong public reaction, which is translated into political action, often in the form of new regulations and costly inspection services.

These costs are either borne by the industry if they are related to the ships and seafarers or by the public in terms of taxes for extra governmental inspection services. It would be much more cost effective if the industry could assess its own risks and take remedial action to prevent accidents. That is the culture that the ISM Code is trying to achieve.

This book describes the background to the ISM Code and, through a major survey, documents how seafarers and managers reacted and responded to their new responsibilities. Not surprisingly, because of the differences in operating philosophies within the industry, there are many differing perceptions and reactions to the Code. These need to be documented if the industry is to understand and learn from past experience.

The book puts the ISM Code into perspective with recent cases which have come before the courts. Courts of law examining commercial claims have to interpret the law, which includes forming judgements based on earlier precedents and the circumstances of the case. Law lords have to be consistent and fair. Now that the ISM Code is a reality there is no doubt that information contained in the safety management system will be asked for as evidence, but nobody should be alarmed by this in the ordinary course of events.

Critical issues arise when the defence of a case may collapse because a vessel was, on the evidence generated from the safety management system, found to be unseaworthy. Clearly, within the context of the claim, defendants will be highly sensitive to the issues. However, in the wider context of shipping operations, the public has a right to expect that ships are seaworthy and so too do the crew. So there is a section in this book covering this subject. The law has not yet built up cases on which to set precedents but there are clear indications about the way such precedents are developing. This, of course, stems from plaintiffs who, not unreasonably, will not pay a claim if they can prove a ship was unseaworthy at the time of the incident.

The key point here is that the law supports good practice. Indeed, it would soon be considered corrupt if it did not. However, there may be a conflict between public policy which requires safer ships and cleaner seas on one hand and the natural instinct of self preservation which pervades a blame culture on the other. The ISM Code with its safety management system will, in the opinion of the authors, become an essential adjunct to the process by which evidence is collected. The practical answer is, of course, to have a good safety management system and then conflict of interest will not arise.

To support this view, it is worth recording that the UK Chamber of Shipping used to publish near-miss accidents together with recent cases from commercial and governmental courts. They had to abandon this practice on the legal grounds that such reports could compromise the liability of the suppliers of information, even though they were without names and confidential.

The Nautical Institute, realising that an essential element of accident prevention had been discontinued, courageously introduced the MARS reports, which are circulated worldwide through *SEAWAYS* and are also available on the Internet. An exciting recent development is that a number of companies are forwarding, in confidential form, incidents which have arisen as a result of feedback from their safety management system. This transformation in attitude is both welcome and correct.

It is essential that people learn from near misses and there can never be any doubt that the law of liability would expressly forbid such practices. However, this example illustrates a deeper framework influencing safety and because of the ISM Code it will change. In future it is likely that judgements will go against companies who inhibit the flow of information, which can lead to accident prevention if such inhibition was found to be material. The proximate cause of an accident, which is the basis of legal decisions, will include an assessment of whether the incident could have been avoided by the ordinary practices of seamen. In other words, custom of the trade and professional advice are, quite correctly, being changed by the ISM Code. In future, companies will come to realise how the Code is changing the values upon which professionals are making their decisions.

This may significantly change the way ships are run and operated in the long term. It was a similar crisis of confidence that revolutionised the way the offshore industry was organised following the fateful *Piper*

Alpha disaster in 1988. Now, the emphasis in the offshore industry is entirely on risk assessment and management systems that acknowledge risks and demonstrate that design and operating practices can be contained within acceptable safety margins. This, of course, includes operatives and their training. The authors believe the shipping industry can learn valuable lessons from this offshore industry experience and the process is explored in later chapters.

Now that the ISM Code is a reality it is time to take stock and consider the way forward. In chapter 13 the author describes how, by sharing common values about safety, a company improves its communication between all participants at a level which generates involvement, stimulates motivation and more open and committed attitudes. Not surprisingly, such attitudes spill over into the whole operating philosophy and practices within a company. When that happens the benefits are truly found and the way to crack the Code is discovered.

This book is written with one further purpose in mind and that is to discriminate between good and bad safety management systems. In the final chapter a model safety management system for a medium sized fleet has been selected. It is not aimed specifically at any particular system, individual or position. It is argued that a managing director or a chief officer will be able to read this book and ask themselves one key question — "How do we compare?"

The next question might be "How can we improve and do we need additional support or assistance with training?" The final section outlines a number of useful training objectives which can be applied collectively, individually in a company or in a training centre.

The aim is to create the culture which reduces accidents, damage, personal injury and lost-time incidents in competitive, commercial ship operations. A lesson that many in our industry still need to learn is that the ISM Code, properly implemented, will not only lead to safer ships and cleaner seas, but also more efficient ships and profitable companies.

Is the ISM Code working?

Fortunately, or unfortunately, the author has worked in the close proximity of lawyers during the last twenty years and therefore feels confident in answering the question in the way only a lawyer could: - … on the one hand it is, but on the other hand it isn't!

5

Hopefully, this book will provide a more illuminating insight into the real answer to that question. At one level the answer really is both yes and no at a deeper level there is a complicated and often complex and conflicting picture — which perhaps should not surprise us when we consider we are dealing with a multinational, multifaceted industry with a globalised labour force. The IMO estimates that some 12,000 ships had to comply by the first deadline, with the second phase of implementation bringing in another 13,000 ships (IMO Briefing 28/6/2002). Accordingly, we should not be too surprised to find a diversity of opinion and experiences.

For one Georgian master it is quite clearly one of the most significant moves forward that has ever happened to our industry:

> *"I think that the implementation of the ISM Code is the most important step in the new millennium for the improvement of safety at sea."*

Whereas, for one British chief engineer, his perception seems quite different:

> *"ISM is the biggest pile of paper eating industrial whitewash that has ever been produced, ships are still being lost or cause incidents, but most will be ISM compliant and the paperwork will be up to date."*

Others are perhaps more philosophical in their reflections such as the master on board a Netherlands Antilles vessel who suggests:

> *"Personally I am more interested in the weather forecast!"*

Or, the maritime university lecturer who suggested:

> *"ISM is a good system badly implemented."*

Can that be reconciled with the claim of one leading, national flag tanker owner who claimed, in unambiguous terms, that:

> *"We are now saving $1,000,000 per ship per year which we directly attribute to our ISM implementation."*

A very important issue to get clear in our minds, even at this early stage, is that the ISM Code is identical, word-for-word, for every ship and every ship operating company everywhere in the world. The reasons why there might be such varied responses and experiences is the real subject of this study.

The only thing that can be said with any degree of certainty about ISM implementation is that there is a very wide-ranging scale of perceptions, level of support, experiences and expectations for this whole ISM phenomenon as the above examples start to illustrate. Much was written in advance of Phase One implementation on 1st

July 1998 about whether the ISM Code was needed and whether it could or would be implemented. Post Phase One implementation much has been written about whether the Code has been implemented. Invariably the authors of those articles and reports tended to be negative in their approach often condemning the ISM Code as a failure or at best a good idea that had been stillborn. Rarely were the authors of those reports the people at the sharp-end of implementation, the seafarers or ship operators. It seemed that their opinions were based on individual bad experiences and not much more than subjective opinions with little or no empirical evidence to confirm that their views were widely held.

Some endeavoured to support their view that the ISM Code was not working by drawing attention to the apparent fact that accidents were still happening at an alarming rate and claims remained at a high level, Port State detentions appeared to be increasing and ships which were in possession of all the relevant paperwork were breaking up, sinking, spilling their oil into the ocean and killing their seafarers. On the other hand there were those who attempted to argue that the Code was working and in their support they groped around desperately looking for something to cling onto — often relying upon a claim by a single P&I Club that they had experienced a reduction in their claims post ISM implementation — which they attributed to the successful implementation of safety management systems by their members. Indeed the Secretary General of IMO himself has made repeated references to the particular report and even published a special briefing on 25.9.2001 entitled IMO welcomes ISM Code study. Those prophets of doom, in the first case, seemed not to appreciate that for every one *Erica* there were many thousands of tankers conducting their trade without incident. That the Port State control authorities had implemented a Concentrated Inspection Campaign and were progressively and aggressively increasing the frequency and intensity of their inspections and that to expect radical changes overnight would be unrealistic and impractical. For those optimists who were almost clutching at straws trying to find something to prove that there was light at the end of the tunnel they failed to recognise that the particular P&I Club which had made the claim was one of the smallest of the clubs, representing less than 2% of the world shipping and thus a minute sample. They also failed to realise that none of the other twelve clubs of the International Group of P&I Clubs, which insure more than 90% of the world deep sea fleet, were making any similar claim nor did any other insurer. Indeed there was a noticeable silence, which should have been apparent to those who had kept an open mind on the subject. Since a lot of good publicity could have been obtained by anyone who could make such a claim then the silence surely tells its own story but the truth behind this is not necessarily as it might appear. The question that should have been asked was …should we realistically have expected to see any significant reduction in accidents and claims, on a global scale, with immediate effect post 1st July, 1998? Realistically the answer would have to be no.

That does not mean however that the ISM Code was doomed from the start only that it would take some time for its benefits to fully manifest themselves across the entire international shipping industry.

However, it was becoming apparent that many ship operating companies, ships and individuals were experiencing significant problems both with the initial implementation and with maintaining their implemented Safety Management Systems (SMS).

The author therefore decided to embark upon a project to collect information and evidence of ISM implementation from the seafarers, operators and other participants in the shipping and related industries and professions. The intention was to try to produce a picture of ISM implementation two and a half years on from the first phase implementation and in the lead up to the second phase implementation.

The project involved a detailed and extensive survey with more than 70,000 questionnaires being distributed worldwide as well as a dedicated Internet website being established at http://www.ismcode.net. The website included details of the research and relevant information about the ISM Code and also allowed the questionnaires to be completed on-line.

Three versions of the questionnaires were produced, as shown in figure 1.1. The first questionnaires were despatched in April 2001 and by the end of November 2001 nearly 3000 completed forms had been returned and the data entered in a Access database. In addition to the answers to the specific questions raised in the questionnaire the respondents were also encouraged to provide narrative comment to share their experience of ISM implementation. Nearly 800 detailed and interesting comments were received and catalogued. All respondents were offered total anonymity. Many chose to exercise that option but many more were quite willing to openly put their names to their comments.

Figure 1.1 — the three survey questionnaires

The author is extremely grateful to all respondents and his only regret is that time and resources did not allow him to respond and thank each contributor personally or to enter into a dialogue particularly when some very genuine and kind offers of help were made. For that he extends his sincere apologies but hopes that the results and conclusions in this book will accurately reflect the views and opinions put forward.

The authors views are expressed extensively throughout this book but, much more importantly, those masters, seafarers, ship operators and other stakeholders are given their opportunity, wherever possible, to have their say in their own words. No apology is made for the numerous quotes which will be included in the text and which will provide much of the substance of this study. If we are to understand how implementation is progressing we need to listen to those who are involved directly in that process we need to hear of the problems experienced as well as the successes achieved. Where mistakes have been and are being made we need to learn from those mistakes and where successes have been achieved let us understand how those successes were achieved and be prepared to emulate those pioneers. It is not suggested that we simply try to copy what others have done, rather we should try to understand the principles and methods adopted and then try to apply them.

It will be noticed from the pictures of the questionnaires that a colour coding was introduced when the questionnaire forms were printed: masters and seafarers in blue; ship operators in green and other stakeholders in red. That colour coding has been carried forward into this book to allow the reader more easily to identify and distinguish between the respective views and results of the masters and seafarers, ship operators and other stakeholders respectively.

9

Anonymity has been maintained for all contributors, even where the individual may have indicated that they were prepared to be identified — except where press reports or conference presentations and similar are being quoted. It was considered sufficient for the purpose of appreciating what the individual had to say merely to define their position in the industry and possibly to identify type of ship or organisation in which they were based and, occasionally, by their nationality. More detailed explanations of the questions in the questionnaire forms and the answers which were provided are addressed in the various sections of this book.

In addition to individuals who returned completed questionnaires or otherwise submitted their personal reflections and observations there were a number of other activities which provided group participation and for which the author is most grateful. For example, Nautical Institute branches around the world held debates involving local branch members to share their experiences and returned minutes or other records of those debates. Nautical colleges in many countries have encouraged their marine students to participate. National ship owners organisations have solicited the views of their members and have encouraged them to participate in the survey. The Baltic International Maritime Council (BIMCO) provided an opportunity for the author to participate in their annual ISM residential course in Copenhagen where the views and comments of the delegates and other speakers were obtained. BIMCO also encouraged their own members to participate and provided a link from the front page of their website to the dedicated ISM site. The author was invited to speak at numerous seminars and conferences which provided further opportunities to meet with and discuss relevant issues with a very wide cross section of people within the industry with important and valuable contributions to make.

Many shipping newspapers and other maritime publications were also extremely helpful in carrying feature articles covering the research and encouraging their readers to participate. All these contacts provided extremely valuable opportunities for the author to widen his own knowledge, understanding and appreciation of the real issues involved in the apparently innocent question: Is ISM implementation working?

Inevitably there must always be some reservations held with regard to the true representation of the data and observations received from questionnaires and a survey of the type undertaken since it could possibly be argued that it is a particular type of person who completes questionnaires and their views may not coincide with the rank and file members of the intended participants and thus may not be a true representation at all. When the various sections of the book are explored it will become apparent that a very wide cross section of views are expressed from those who are vehemently against ISM to those who are 100% supporters, with many shades of grey in between. There are seafarers of all ranks and many nationalities. It is the view of the author that the results as set out in the following sections of this book do provide an accurate reflection of the varied views of a significant number of those at sea and those ashore on the question of ISM implementation during the year 2001.

Whilst it may appear unfair to inflict a conclusion on the reader at the beginning of a work such as this it is believed that on this occasion it is important if objectivity is to be maintained and a serious bout of depression is to be avoided. A very important point to take hold of at this stage and to carry forward as we proceed into the substance of the book itself is that the ISM Code can work and is working very satisfactorily in a number of companies. This will hopefully be demonstrated in due course. There will, though, be many negative issues coming to light in the sections ahead. It is important that we confront those issues but, at the same time, keep in mind the knowledge of the possibility that a properly implemented SMS can work and is working on board ships today.

As we now start to consider, in more depth, some of the findings from the survey it is important to introduce some consistency into our understanding and use of particular terms. A number of the expressions used do have quite specific meanings and are defined within the relevant regulations. It is therefore appropriate, perhaps, to set out those definitions as contained in SOLAS Chapter IX, resolution 741(18) / MSC.104(73) and resolution 788(19) / A.913(22):

- **Accident** means incidents involving injury or damage to life, the environment, the ship or its cargo.

- **Hazardous occurrences** are situations which could have led to an accident if they had developed further (i.e. near miss situations).

- **Non-conformity** means an observed situation where objective evidence indicates the non-fulfilment of a specified requirement.

Five further definitions, taken from the ICS/ISF Guidelines on the application of the ISM Code, are also useful:

- **Internal SMS Audit** is a systematic and independent verification process carried out by the Company as part of its management function to determine whether the SMS activities and related results are in compliance with the SMS.
- **Objective evidence** means quantitative or qualitative information, records or statements of fact pertaining to safety or to the existence and implementation of a SMS element, which is based on observation, measurement or test and which can be verified.
- **Observation** means a statement of fact made during a safety management audit and sustained by objective evidence.
- **Safety management audit** means a systematic and independent examination to determine whether the SMS activities and related results comply with planned arrangements and whether these arrangements are implemented effectively and are suitable to achieve objectives.
- **Verify** means to investigate and confirm that an activity or operation is in accordance with a specified requirement. (Examples of verification activities includes inspections, tests and operational checks on ships and their equipment prior to departing port, at sea or before entering port or closing with the land. A system audit is also an example of a verification activity.)

Chapter 2

BACKGROUND TO THE ISM CODE

Introduction

Commercial shipping is very old — certainly there is evidence of trading ships existing more than 2,500 years before the Christian era. To a very large extent the shipping industry has been self-regulating throughout this very long history. Traditionally, ships were subject to the laws, rules and regulations of the Flag State to which they belonged. They were also obliged to comply with the local laws of the countries they visited. During the period from the early 17th century to the latter part of the 20th century it was quite true that 'Britannia ruled the waves'. The merchant fleet of Great Britain dominated international trade — along with the fleets of other colonial powers such as France, Holland, Spain and Portugal. The merchant marine was a vital factor in the development of international trade, the expansion of the Empire and the prosperity of the nation — as well as a number of individual businessmen. Anyone who had sufficient funds could purchase a vessel and enter the business of shipping.

Britain became a centre for the development of maritime law and marine insurance and since it was so influential in international trade it was very much British rules that applied internationally. Ships were often armed with canon and carried marines — they were run very much along the disciplined lines of the Royal Navy. Against this background, ship owners were allowed to run their companies with little supervision by government — provided they obeyed the law.

The origins of international maritime conventions can be traced to the late 19th century and early 20th century. However it was not until the years following the Second World War, with the formation of the United Nations, that commercial shipping started to become more regulated on an international level. The United Nations Convention on the Law of the Sea, 1982 (UNCLOS) established the general rights and obligations of the Flag State. Within the United Nations two specialised agencies deal with maritime affairs; the International Maritime Organization (IMO) and the International Labour Organization (ILO) and they have a responsibility for devising and developing conventions and guidelines under which ships can be regulated. In general, matters concerning safety at sea, pollution prevention and training of seafarers are dealt with by IMO, whereas the ILO deals with matters concerning working and living conditions at sea. While IMO and ILO set the international regulatory framework for ships, each member state bears the responsibility for enforcing the international conventions it has ratified on the ships flying its flag. However, the industry was still allowed to regulate itself within the confines of these conventions once ratified by their flag states as well as other elements of the domestic law of that country.

Up until the period following the Second World War almost all merchant ships flew their own national flag. However, led by ship owners from the United States, an increasing number re-registered their ships and companies in countries where application of the rules and regulations was a little more relaxed or provided tax advantages. These were the so called Flags of Convenience (FOCs) or Open Registries. From inception, FOCs were perceived by many as an opportunity to lower the very high, but costly, standards that had been maintained on board national flag fleets. Even so, during those post war years there were fleets to rebuild and trades to re-establish which meant that merchant ships were fully employed in helping to bring the world back to normality.

In the late 1960s, when the author first went to sea, a 15,000 ton general cargo ship would typically have a complement of 65 officers and crew on board. Most officers, if not crew members, would be on long term company contracts and it was not at all unusual for a seafarer to remain with the same family shipping company for his entire career. The loyalty, which was reciprocal between employer and employee, was very strong. Ships were well run with good, well-qualified and motivated seafarers.

During the second half of the 1980s and early 1990s, there seemed to be an explosion of maritime accidents and claims generally, with a significant number of high profile major incidents, some of which appear in the following list:

1987 *Herald of Free Enterprise* capsized off Zeebrugge. 190 people lost their lives.

1987 *Dona Paz* ferry collided with a tanker in the Philippines — an estimated 4,386 people were killed.

1989 *Exxon Valdez* ran aground off the coast of Alaska, spilling 37,000 tonnes of oil and causing extensive environmental damage. Final claims level may possibly exceed US$10 billion.

1990 *Scandinavian Star* ferry disaster. 158 people died.

1991 *Agip Abruzzo* with 80,000 tonnes of light crude on board was in collision with the ro-ro ferry *Moby Prince* off Livorno, Italy. Fire and pollution occurred and 143 persons died.

1991 *Haven* fire and explosion off Genoa. Claims arose in excess of US$700 million.

1991 *Salem Express* Egyptian ferry struck a reef and sank. 470 people were killed.

1992 *Aegean Sea* broke in two off La Coruna, Spain. Extensive pollution occurred. Claims approached US$200 million.

1993 *Braer* driven onto the Shetland Islands, causing widespread pollution. Claims in region of US$200 million were presented. The Donaldson Inquiry was set up in the UK.

1994 *Estonia* ro-ro passenger ferry sank after the bow door fell off during heavy weather at sea. 852 people lost their lives.

1996 *Sea Empress* caused major oil pollution off Milford Haven, UK.

Problem — an accident and claims explosion

During the period 1987 to 1990, P&I insurance claims and consequently the cost of P&I insurance rose on average by 200-400%. It is understood that a similar phenomenon was also experienced with hull and machinery claims and premiums. It became apparent that the international shipping industry was perhaps no longer capable of regulating itself and action was needed to reverse the downward spiral of maritime calamity. It was against the background of this catastrophic situation that the author first became involved in looking at the problem of maritime accidents and to consider what he could contribute to help remedy the situation. By the late 1980s alarm bells were ringing in many quarters. The shipping industry seemed to be in a disastrous state and few could provide any rational explanation as to what was going wrong. Numerous investigations and reports were commissioned by government agencies and by industry to try and throw some light on the problem. In 1988, for example, the UK Department of Transport funded research carried out by the Tavistock Institute which led to the report 'The Human Element in Shipping Casualties' (Department of Transport, 1991). The conclusions of that report were taken to the IMO by the then Surveyor General's Organisation (SGO) whose role was later taken over by the Marine Safety Agency (MSA).

15

The world's largest P&I Club is the United Kingdom Mutual Steamship Assurance Association (Bermuda) Limited, who provide P&I insurance to approximately 25% of the world fleet. In 1991, through its managers Thomas Miller P&I, the club issued its first 'Analysis of Major Claims' (the UK Club, 1991). In 1992 the House of Lords Select Committee on Science and Technology, under the then chairmanship of Lord Carver, issued its report on the Safety Aspects of Ship Design and Technology (House of Lords, 1992).

Cause — human error

A common factor appeared in each report. Basically these accidents and incidents were arising primarily as a result of human failings. For example, in the 'Human Element in Shipping Casualties' report, it was stated that the human element was found to be causative in over 90% of collisions and groundings and over 75% of contacts and fires/explosions. The UK Club report concluded that human error accounted for 58% of all claims and the House of Lords/Carver Report concluded at Section 4.2 that "... It is received wisdom that four out of five ship casualties — 80% — are due to 'human error ....'".

On reflection, these conclusions should not have come as any great surprise. Whilst statistical data is probably not available, it is suggested that 'human error' or 'human factors' (or whatever other title one wishes to label the phenomenon) have been responsible for almost all maritime accidents throughout history and that the figure is probably much closer to 100% The deciding factor depends on where the investigator/researcher stops in tracing a particular causal chain for any particular accident. Part of the problem was the fact that there were more accidents and claims — costing more in terms of lives, environmental damage and money than ever before — and the situation seemed to be getting worse. The problem was much more complicated than that though.

At the end of the day the real problem was economics. Almost the whole of the shipping industry was in deep economic recession. This had major knock-on effects as the industry tried to survive in such very difficult financial times. The nature of the economic problem is quite easy to understand — it was the most basic of economic principles — the law of supply and demand. Basically there was a surplus of ships for the number and volume of cargoes to be carried. Many traditional shipowners sold their ships and got out of the industry.

Others looked for ways to cut operating costs to levels that might allow them at least to break even with the very low freight and charter-hire rates that they were being offered. Flying the national flag often involved restrictive practices with regard to labour laws and such things as compliance with safety related legislation. A flood of ship operators therefore de-registered and hoisted strange flags of convenience on their ships — registering the owning Company as a 'one ship company' with a brass nameplate on a doorway in some tax friendly country.

The wage bill was an obvious and immediate target, both in offices ashore and with seagoing staff. Marine superintendents ashore, who had provided a vital link between ship and shore, found themselves redundant. Safety, training and personnel officers ceased to exist, almost overnight. Legal and claims department staff found themselves expendable and assistants within the various departments were looking for other employment. Those who were left — the operations manager and technical superintendents — had to try to continue doing their own jobs and also the jobs of all those who had been casualties. On board ship the situation was even worse!

Traditional seafarers from the UK, Scandinavia, Northern Europe and the Mediterranean were perceived to be too expensive. Cheaper labour supplies were identified in developing nations — particularly in South East Asia. In a very short period of time, highly skilled and well qualified traditional seafarers were displaced by seafarers having little basic education and even less maritime education and training. Of equal concern was the fact that actual numbers of personnel on board were being reduced significantly, compounding the problem. Quite typically the number of officers and crew were being reduced by between one half and two thirds.

People were not the only cost cutting target though. Ships themselves were built to have a typical trading life expectancy of about 15 to 20 years, after which they would be scrapped and replaced with new buildings. New buildings were prohibitively expensive and so ships were being traded well beyond their natural life. Simultaneously, maintenance budgets were being slashed. Without vital maintenance the condition of ships quickly deteriorated, resulting in an increased risk to people, the cargo being carried and indeed the ship itself. To compound this problem of older ships receiving less and less maintenance was the apparent relaxation of standards by the classification societies. The societies had performed two very important roles for many, many years. Firstly they carefully monitored the construction and maintenance of ships which provided a type of risk assessment and assurance/guarantee for the hull and machinery and P&I insurers. Secondly, acting on behalf of various flag states, they monitored and assessed compliance by the shipping company with a whole range of important safety-related legislation. The societies, however are financed by, and consequently their activities potentially influenced by, the ship operating industry.

There were other factors too which contributed to a cocktail of disasters for the shipping industry. The key factors, however, were people and management systems. The core issues were identified by the Secretary-General to IMO in a Briefing dated 28.6.2002 – 'Shipping enters the ISM Code era with second phase of implementation' – when Mr. O'Neil said: "Previously, IMO's attempts to improve shipping safety and to prevent pollution from ships had been largely directed at improving the hardware of shipping — for example, the construction of ships and their equipment. The ISM Code, by comparison, concentrates on the way shipping companies are run." He continued "this is important, because we know that human factors account for most accidents at sea — and that many of them can ultimately be traced to management. The Code is helping to raise management standards and practices and thereby reduce accidents and save lives'.

Solution – management systems

Efforts to address the problem had already started a little earlier. In July, 1986, for example, following publication of the report on the loss of the *MV Grainville*, the British Government issued M Notice 1188. This was subsequently updated and superseded by M Notice 1424 in August 1990, entitled 'Good Ship Management'. This commended the publication entitled 'Code of Good Management Practice in Safe Ship Operations' issued by the International Chamber of Shipping and the International Shipping Federation.

The tragic loss of the *MV Herald of Free Enterprise* in March, 1987 resulted in the introduction of the Merchant Shipping (Operations Book) Regulations 1988 which were laid before Parliament and came into force in December that year. These regulations are applicable to all UK passenger ships on short sea trade (Class II and IIA) and were developed around the two central tenets of:

- All such ships having to carry an operations book containing instructions and information for safe and efficient operation.

- Owners being required to nominate a person (known as the Designated Person) to oversee the operation of their ships and to ensure that proper provisions are made so that the requirements of the operations book are complied with.

M Notice 1353 was issued in October 1988, giving detailed guidance on how compliance with these regulations could be achieved. Attention was also drawn to M Notice 1188 and the Code of Good Management Practice in Safe Ship Operations. At the 57th session of the IMO Maritime Safety Committee (MSC) in May, 1989, the UK delegation pressed, unsuccessfully at that time, for the draft guidelines contained in MSC 56/WP.4 (this working paper was ultimately adopted at the 16th Assembly, in October, 1989, as resolution A.647(16) and is the forerunner to the ISM Code) to include the two principles upon which the Merchant Shipping (Operations Book) Regulations 198/8 were founded.

Further impetus was given to the need for these amendments to SOLAS when fire swept through the Norwegian passenger/car ferry *MV Scandinavian Star* in April, 1990 with the loss of 158 lives. That tragedy initiated the action within the IMO that resulted in the inclusion of paragraphs '4.7 Designated Person Ashore' and '4.8 Operations documentation in resolution A680(17)' which was adopted on 6th November, 1991 revoking resolution A.647(16).

At the 18th session of the IMO Assembly on 4th November, 1993, resolution A.741(18) was formally adopted. This revoked resolution A.680(17) and constitutes verbatim the International Management Code for the Safe Operation of Ships and for Pollution Prevention (International Safety Management (ISM) Code). This was incorporated on 19th May, 1994 into the SOLAS Convention 1974 as Chapter IX — entitled 'Management for the Safe Operation of Ships', making compliance with the Code mandatory in states that are signatories to SOLAS for various classes as follows:

- Ro-ro passenger ferries operating between ports in the European Union by 1st July, 1996 — pursuant to a regulation of the Council of the European Union.
- Passenger ships including high speed craft, not later than 1 July, 1998.
- Oil tankers, chemical tankers, gas carriers, bulk carriers and cargo high speed craft of 500 gross tonnage and upwards, not later than 1st July, 1998.

- Other cargo ships and mobile offshore drilling units of 500 gross tonnage and upwards, not later than 1st July, 2002.

(The Code does not apply to government-operated ships used for non-commercial purposes).

Alongside Resolution A.741(18) – The International Safety Management (ISM) Code – the IMO also developed Resolution A.788(19) – Guidelines on Implementation of the International Safety Management (ISM) Code by Administrations – which was adopted on 23rd November 1995. Resolution A.788(19) was intended to provide the Flag State Administration with a set of outline guidelines which they could use when looking at the SMSs of the shipping companies and ships under their flag to verify compliance with the ISM Code and the issuance of the DOCs and SMCs. The intention was to introduce some uniformity and consistency into these processes on an international level. It is important to understand however that these were guidelines only, without any mandatory or compulsory status. It would appear that the intention of the authors and architects of the Code was that the two resolutions would sit side by side and complement each other.

Resolution A.788(19) did appear to be taken on board by many administrations but also led to a certain amount of confusion. The opportunity was therefore taken at the December 2000 meeting of the IMO Maritime Safety Committee meeting — MSC.99(73) — to amend the text of the ISM Code (Resolution A.741(18)) specifically to include a number of provisions from Resolution A.788(19) and to replace that Resolution with a new draft A.913(22). The new Code and amended Resolution came into full force to coincide with Phase Two implementation on 1st July 2002.

The original intention of IMO was that the Flag State Administrations would be the bodies undertaking the verification and certification on board the ships flying their national flag. A limited number of administrations did undertake this work but the IMO recognised the increasing dominance and influence of the Flags of Convenience (FOCs) and the fact that many of the FOCs had very limited infrastructure actually to undertake this task. Accordingly, IMO built into the text of SOLAS Chapter IX and Resolution A.733(19) flexibility to allow administrations to delegate the actual task, but not the responsibility, of the verification and certification to Recognised Organisations (ROs) or to other Administrations. Almost all FOCs and many national administrations, delegated to the classification societies and a small number of independent consultants. Many of these same societies and consultants had also set up consultancy companies in which they were selling their expertise to ship operators to set up, develop and write their safety management systems. There were and still are many in the industry who considered this dual role runs the risk of a very serious conflict situation arising. Those who were setting up the systems were then examining their own efforts and issuing certificates — many questioned the objectivity and indeed the ethics of such a practice. The implications and significance of this somewhat incestuous situation were to figure in the findings of this research.

Chapter 3

THE PHILOSOPHY OF THE ISM CODE

Introduction

The stated objectives of the ISM Code are initially set out in the first paragraph of the preamble to the Code which provides:

> "The purpose of this Code is to provide an international standard for the safe management and operation of ships and for pollution prevention".
>
> This initial statement is expanded and the objectives are set out in full in section 3.2.
>
> ### 3.2 Objectives
>
> 3.2.1   The objectives of the Code are to ensure safety at sea, prevention of human injury or loss of life, and avoidance of damage to the environment, in particular to the marine environment and to property.
>
> 3.2.2   Safety management objectives of the Company should, *inter alia*:
>
> .1 Provide for safe practices in ship operation and a safe working environment.
>
> .2 Establish safeguards against all identified risks.
>
> .3 Continuously improve safety management skills of personnel ashore and board ships, including preparing for emergencies related both to safety and environmental protection.
>
> 3.2.3   The safety management system should ensure:
>
> .1 Compliance with mandatory rules and regulations.
>
> .2 That applicable Codes, guidelines and standards recommended by the organisation, Administrations, classification societies and maritime industry organisations are taken into account.

A change of perspective on rules and regulations

The author agrees with the points raised in the Intertanko document (Tatham) when they discuss the factors which led up to the Code's objectives being formulated. At paragraph 3.2 they explain

"...The Code was produced in response to potential pressure, following a number of high profile incidents, for the shipping industry to clean up its act, the perception being that the existing rules and regulations were not in themselves sufficient to ensure a real diminution in the number of shipping casualties — in particular it was felt there was a need to reduce the scope for human error by imposing an industry standard of good management ...".

The whole intention of the ISM Code has been summarised so well by Lord Donaldson when he said

"... In the short and medium term, it (the ISM Code) is designed to discover and eliminate substandard ships, together with substandard owners and managers, not to mention many others who contribute to their survival and, in some cases, prosperity ...".

He continues

"... In the longer term its destination is to discover new and improved methods of ship operation, management and regulation which will produce a safety record more akin to that of the aviation industry. But, as I readily admit, that is very much for the future ..."

The ISM Code is concerned with procedures whereby the safety and pollution prevention aspects of a ship are managed, both ashore and on board, rather than laying down specific rules as to the technical condition of the ship itself. There are of course numerous sets of rules, regulations and conventions dealing with technical issues which a shipowner will need to comply with. Indeed the full significance of section 1.2.3 can be easily overlooked and certainly underestimated. The Code is not necessarily introducing any new sets of rules and regulations, but rather provides a requirement that the SMS should be structured such that it can check and verify compliance with all the various existing rules and regulations. Such rules and regulations would include, by way of examples, Load Line Regulations, Radio Regulations, Collision Regulations, MARPOL, the other chapters of SOLAS, Classification Society Rules, STCW and a host more. A misunderstanding which seems frequently to arise in the minds of many people is to think that the ISM Code has suddenly become an all encompassing, all embracing piece of legislation, incorporating all these various sets of rules, regulations, conventions and legislation. The Code does not incorporate them at all although it is a breach of these regulations that principally gives rise to exposure. What the ISM Code does - as stated in its objectives - is to make a requirement that the safety management system will provide procedures by which a company can check that it does comply with the various rules and regulations and procedures to check and verify that they continue to comply. Another important point to understand is that these procedures must be documented and recorded. In this regard it is perhaps worth noting the contents of section 2.3.2 of Resolution A.913(22) which states:

"All records having the potential to facilitate verification of compliance with the ISM Code should be open to scrutiny during an examination. For this purpose the Administration should ensure that the Company provides auditors with statutory and classification records relevant to the actions taken by the Company to ensure that compliance with mandatory rules and regulations is maintained. In this regard the records may be examined to substantiate their authenticity and veracity".

Flexibility within the system

To understand fully how the stated objectives should be interpreted, it is very important to understand the significance of paragraphs 4 and 5 of the Preamble. These read:

> **Preamble**
>
> 4. Recognising that no two shipping companies or ship owners are the same and that ships operate under a wide range of different conditions, the Code is based on general principles and objectives.
>
> 5. The Code is expressed in broad terms so that it can have a widespread application. Clearly, different levels of management, whether shore based or at sea, will require varying levels of knowledge and awareness of the items outlined.

It is of considerable credit to the draftsmen of the Code that they intentionally drafted it in such broad terms. Having achieved that, the general principles which are set down are of widespread application to all types of ship and owner. The very general terms in which the ISM Code is written do need to be understood within the context of the safety case as proposed by Lord Carver. This was put very well by Lord Donaldson when he said:

"... what the ISM Code seeks to do is to superimpose a safety case regime which is regulatory in the sense that it is compulsory and is intended to be fully enforceable, whilst being specific only in its general requirements ..."

The other important point to recognise is that the Code does not anticipate all levels of management, either ashore or on board, to necessarily have the same levels of knowledge and awareness of safety and environmental issues.

If one needed an explanation as to why it was considered necessary to look afresh at the way maritime safety was managed which moved away from a regulatory framework to one based upon a safety culture, one would only need to reflect upon the sobering and appalling rate of bulk carrier losses. During the period 1980—1997, there were reportedly 167 bulk carriers lost and, what is infinitely worse and unacceptable, is that 1,352 lives were lost on those vessels.

Safety and environmental protection policy

The ISM Code requires the Company to provide, in clear and concise terms, a statement describing what its aims and intentions are with regard to its SMS along with outline details of how those aims and objectives are to be achieved. The requirements of the Code are set out in Section 2:

> ### 2 Safety and Environmental Protection Policy
>
> 2.1 The Company should establish a safety and environmental protection policy which describes how the objectives given in paragraph 1.2 will be achieved.
>
> 2.2 The Company should ensure that the policy is implemented and maintained at all levels of the organisation both, ship-based and shore-based.

Implications for the Company

The significance and importance of the Safety and Environmental Protection Policy should not be underestimated — for it may come under very close scrutiny if a major incident occurs.

The policy is basically a statement by the Company to the Company and all its employees whether ashore or on board. It is recommended that the policy should be signed by the chief executive or similar head of the organisation to demonstrate the commitment from top level management. It should also be regularly reviewed.

It is very important that the statement in the policy really does voice the intentions, aspirations and commitment of that particular company rather than some eloquent prose provided by some so-called consultant selling off-the-shelf safety management systems.

From a practical point of view such a policy would be needed in order for the Company to have credibility with its personnel with regard to its commitment to safety and the protection of the environment. From a pragmatic point of view it should be anticipated that if an incident did arise involving the Company then the press and media would certainly scrutinise the policy most carefully and use it to criticise, ridicule or condemn the Company.

For different, but related reasons, the courts would also look very carefully at the policy statement as well as the historical track record of the Company in light of the policy statement. Some flowery, standard worded bought off-the-shelf policy statement could cause considerable embarrassment indeed to a company.

The safety management system

It would not be an overstatement to claim that what the ISM Code is all about is the development and implementation of a Safety Management System (SMS).

The responsibility is upon the owning or operating company to develop, implement and maintain not only a written but a living, dynamic, safety management system covering a whole range of safety, environmental and related matters. The functional requirements are listed in Section 1.4 of the Code and are then explored in more detail throughout the rest of the sections of the Code.

1.4 Functional requirements for a safety management system

Every company should develop, implement and maintain a safety management system (SMS) which includes the following functional requirements:

.1 A safety and environmental protection policy.

.2 Instructions and procedures to ensure safe operation of ships and protection of the environment in compliance with relevant international and Flag State legislation.

.3 Defined levels of authority and lines of communication between, and amongst, shore and shipboard personnel.

.4 Procedures for reporting accidents and non-conformities within the provisions of this Code.

.5 Procedures to prepare for and respond to emergency situations.

.6 Procedures for internal audits and management reviews.

Implications of the functional requirements

The functional requirements as set out in Section 1.4 of the Code are really just the main chapter headings of the ISM Code — the named organs, as it were, of the SMS. They are not intended to be an exhaustive list but rather represent the minimum requirements of an acceptable safety management system.

It is the structure, implementation and working of the SMS which will be the real deciding factor of whether or not a company is complying with the ISM Code.

Getting inside safety management systems

The greatest strength of the ISM Code is simultaneously its greatest weakness. It is arguably the single most important and influential piece of maritime legislation ever to have been enacted on an international scale, yet the Code itself is set out in 16 short sections on fifteen pages in an A5 size booklet. Its greatest strength lies not only in its simplicity but in its flexibility; the original draughtsmen on the International Maritime Organization (IMO) working party produced a set of very general principles and objectives, with a widespread application, which could be interpreted and implemented by each individual shipping company or shipowner or ship as may be most appropriate to the way in which that operator managed its company and its ships. The intention was that compliance would be individual and there would be considerable latitude and freedom in producing a SMS which would be most suitable for that individual operator. Of course each individual company and ship would be bound by those general principles and objectives, as well as all the other pieces of international and national legislation, industry and insurance rules and regulations and other contractual terms. The point is though that the ISM Code allows each operator to do it their own way.

The other side of that coin though produces what is perhaps the greatest weakness of the ISM Code; with so many individual interpretations and applications of the Code how is it possible to objectively measure compliance? This aspect needs splitting into at least two component parts to demonstrate the complexity of the issue under consideration — verification of initial compliance, and post implementation. We must consider not only verification of compliance as far as setting up the initial SMS is concerned — for which the Documents of Compliance (DOCs) would be given to the operating company and the Safety Management Certificates (SMCs) to the ship — but much more difficult to measure is the subsequent monitoring of successful implementation. Complicating the matter even further, the determination as to whether an individual ship operator and/or ship complied and had implemented an acceptable SMS was made the responsibility of the individual Flag State Administrations, i.e. the government of the flag which is flown by the particular ship. Whilst IMO promulgated Guidelines on Implementation of the International Safety Management (ISM) Code by Administrations in 1995 by way of Resolution A.788, it was still left to each Administration to decide how closely, if at all, they would follow those guidelines. Complicating the issue yet further, flag states can delegate the task, but not the responsibility, to third parties to undertake the verification process and implementation on their behalf. IMO Resolution A.739(18) Guidelines for the Authorisation of Organisations Acting on Behalf of the Administration became mandatory by virtue of the new SOLAS chapter IX and Resolution A.740(18) — Interim Guidelines to Assist Flag States — whereby Administrations can authorise organisations to issue DOCs and SMCs on their behalf. Those other organisations became referred to as Recognised Organisations (ROs) and, understandably, they would also have their own interpretations to apply. As a vessel moves around the world, from country to country, it is quite likely that it will be visited by inspectors from the local Port State Control (PSC), or similar body, who may wish to satisfy themselves that the ship complies with the requirements of the ISM Code. Clearly each individual PSC will have its own benchmark against which it measures acceptable compliance and that is possibly based on the criteria decided by the government of that country when considering ISM from its position as Flag State Administration. Compliance of any individual SMS could therefore be determined by the opinion of the ship operator, or more correctly the Company, the Flag State Administration or RO and possibly PSC.

Experience suggests that objective standards have varied widely from company to company, from flag to flag and from PSC to PSC. It has been suggested that some companies have switched flag or in some cases the RO in order to obtain more easily the DOCs and SMCs and it has even been reported that DOCs and SMCs can be bought from certain Administrations with little or no verification having taken place at all!

How does one ascertain and measure compliance objectively with an international code when there are so many different subjective interpretations actually determining compliance? Establishing that the DOC and SMC exist merely confirms that the particular Administration deemed it appropriate to issue those particular certificates. Whether that is the same as saying that the SMS would satisfy any other Administration's verification criteria is, perhaps, another matter. In some cases those certificates may be nothing more than worthless pieces of paper.

Of course there may be situations which manifest themselves externally to the SMS which might provide an indicator with regards to compliance, or non-compliance as the case may be. If, for example, defective, badly maintained or missing safety equipment is evident, this would suggest not only a failure to comply with Load Line Regulations or other sections of SOLAS but also a failure of the management system which should have had provisions with regard to planned maintenance, inspections, testing and auditing to check that those systems were working and effective. Such deficiencies with equipment may result in the vessel being detained by PSC.

However, such a manifestation may only indicate, or highlight, that there appear to be deficiencies in certain parts of the SMS — albeit very important areas — and it maybe that other parts of the SMS are working very well.

Another external indicator might be the record of accidents, incidents and claims either for the Company as a whole or the individual ship. If the trend is clearly downwards then it may suggest that the benefits are being reaped but if the trend is in the other direction then maybe this would indicate that some further adjustment may be needed with the SMS.

We must be careful though not to jump too quickly to what may appear obvious conclusions. However tempting it may be, we should not underestimate what is involved in the implementation of the ISM Code. The philosophy underpinning the ISM Code will require all the seafarers around the world, their employer, their legislators and their regulators to change their whole approach to how they undertake their job. For some this will be a much bigger change than it will for others. Such a change will not happen overnight and for some it may take a long time. There may also be other, less obvious, external factors occurring which may be distorting the overall picture. One such factor could be the increasing tendency of individuals to litigate and pursue claims with an expectation of a payout regardless of merits or genuine claimants with an expectation of a very high level of settlement. We are living in a more litigious society and this factor is having a marked effect on insurance claims.

We can therefore consider evidence derived from, say, PSC inspections/detentions, or class records or insurers claims statistics as providing an indicator with regard to ISM compliance but with some serious reservations as to its full significance. Some of this evidence will be explored and its significance considered in much more detail in Chapter 4.

If it is accepted that each SMS is different and individual and hence direct comparisons with a master blueprint not possible and that the manifestation of external problems and accident/claims statistics are at best mere indicators that the SMS may or may not be having a desired effect, we need to ask ourselves if there is any other way in which the working SMS can be observed or measured.

The ISM Code was conceived in response to the recognition that the vast majority of all accidents and incidents at sea can be attributable to human factors — some might use the term human error. Underpinning the ISM Code and the safety management system is the idea of a safety culture being developed amongst those in the Company ashore and amongst those working on board ship. It is suggested, and it is the basis upon which the research project behind this book was based, that the only way to observe if the SMS has been successfully implemented is to get inside the SMS, into the safety culture to seek the views and perceptions of those directly involved with the implementation.

By combining the observation of the SMS from the inside with the data available relating to external manifestations we can produce a reasonably clear big picture of the current status of ISM implementation. We can identify where problems exist, the extent of those problems, what solutions might be available and what remedies may already have been found or what may need to be prescribed.

Without the creation of this big picture — where progress can be demonstrated — there must be a risk that the greatest weakness of the ISM Code, its deliberate avoidance of having detailed, standardised sets of rules and regulations as a benchmark against which compliance can be more easily measured (as some may claim) may be used by some to impose those detailed rules and regulations and virtually take away the latitude and freedom which each individual ship operator currently enjoys.

There remains though, perhaps, a legitimate questions to ask: 'With such a diversity and proliferation of safety management systems in existence, how will it be possible to be sure that any particular system has been adequately created, developed and implemented effectively? If there is no model or ideal SMS against which to measure other systems, how can we assess which is a good system, which is a bad system, which system is efficient and which system is inefficient? These are fair and reasonable questions to ask. The answer can perhaps be best explained by way of an analogy: imagine trying to describe an elephant to someone who had never seen such an animal. This would not be an easy task and the mental picture the other person may form

of this strange animal may actually be far different from the real thing. However, once that person sees an elephant for the first time there is no mistake in their mind. There is no doubt, they know exactly what it is that they have just encountered. Subsequent encounters with similar animals makes them easily recognisable but that person would have the same difficulty describing the elephant to anyone else who had not been so fortunate to have had the first hand direct experience of an encounter! In the same way when a good, working, dynamic, living SMS is encountered there may be some pleasant surprises but there will be few doubts in the mind of the individual who is having the experience as to what it is that they have come across.

One of the first and perhaps most influential experiences of this nature encountered by the author was a visit to a medium sized tanker operator a few years ago. The managing director of the Company had specifically asked to meet with his visitor on arrival. After a brief exchange of pleasantries the MD, clearly very pleased and proud about something, launched into a somewhat disjointed description of how well his company was now doing. Accidents and claims were on their way down, morale was up and, he confided with a smile as wide as a Cheshire cat, profits were up! He unreservedly attributed the good times being experienced by the Company to the ISM Code — to quote him, almost verbatim,

"I wish we had done this ISM thing years ago".

Clearly there was one person in the Company who was most upbeat about the ISM Code and it was difficult not to be affected and impressed by the MD's open enthusiasm. That was only the beginning though. He had organised a meeting with his in-house lawyer, accountant, technical managers, operation managers, safety managers and others. The MD's support for ISM was very clearly reflected in each of the other executives and managers at that meeting. The ISM bug was clearly contagious in that company. They had seen the advantages of making a total and unequivocal commitment to safety and the Company was now seeing the benefits. The support and leadership from the very top was unambiguous and everyone was on the same song sheet and at the same party! Without the need for a close examination of the procedures manuals or the records of reporting hazardous occurrences or inspecting any of the detail, it was clear that what was being experienced here was the shore based side of a very successfully implemented safety management system. It is only by personal experience of the aura and the buzz in such an environment that anyone could really grasp the living and dynamic nature of such a management system. Unfortunately, the author has not yet had the opportunity of visiting any ships of that company but would be very surprised if the enthusiasm of the MD and the management ashore had not had a most profound and positive effect on board.

The problem was, therefore, how to get inside as many safety management systems as possible to see how they were structured and functioning. To achieve that end would require soliciting the views of a very wide range of individuals who were directly or indirectly involved in the running of those systems. Logistically it would not be feasible to meet sufficient individuals face to face to discuss their views and therefore other arrangements would have to be put in place to canvas the wide range of views needed. A major survey of the entire international shipping industry and related professions was conceived which would involve a number of different methods of approach.

It was recognised at an early stage that this research project would be the first major study into the implementation since the Phase One compliance deadline of 1st July 1998, though there had been a number of limited surveys on related issues. In an attempt to apply rigorous objectivity and sound research practices to the survey methodology the author decided to conduct the research under the guidance and supervision of a university. To this end he registered the project with the National Centre for Work Based Partnerships at Middlesex University, where the results of the research would be submitted towards a doctorate in professional studies. The timing of the project was such that it would coincide with the run-up to the Phase Two implementation deadline of 1st July 2002.

Chapter 4

WHAT EVIDENCE IS THERE TO DEMONSTRATE THAT THE ISM CODE IS WORKING?

Introduction

An observation received from a British chief engineer probably said what was on the minds of many in the industry:

"The implementation of the ISM has been expensive for all companies. The workload for officers on board has increased considerably due to its implementation (and preparation in our case). When will we be able to see statistics to justify its implementation?

We are now nearly three years since the first phase compliance — where can we find statistics on its impact, particularly on British-registered vessels? Those of us still trying to achieve implementation/compliance would like to be assured it isn't just a paperwork exercise."

This is not an unfair or unreasonable request. Unfortunately, however, it is probably going to be a few years yet before any global or industry wide figure can demonstrate what the chief engineer really wants to see.

The Secretary-General of IMO seemed quite confident in his address at the 25th annual World Maritime Day presentation when Mr. O'Neil drew attention to the continuing decline in lives and ships lost at sea and to concurrent reductions in maritime pollution. Koji Sekimizu, representing IMO at the IMarEST Conference probably disclosed the source of that claim as Lloyd's World Casualty Statistics which suggest that in 1995, three ships were lost for every thousand in the world fleet but in 2000, the equivalent figure was 1·9 for every thousand.

Most of us like to see facts and figures — we feel comfortable with them and reassured by them. Hard and fast figures provide a firm foundation upon which we can base our knowledge. The actual situation, however — as far as ISM implementation is concerned — appears to be complex and very complicated.

There are at least three obvious places to look for the statistical evidence that might demonstrate whether the ISM Code is working, or at least is starting to have an effect:

1   The marine insurance sectors — e.g. hull and machinery insurers, the P&I Clubs and cargo insurers. Their records of claims and claim trends should provide an indication.

34

2	The Port State control MOUs and their secretariats. Their records of inspections and detentions should show whether the number of ISM related deficiencies noted are increasing or decreasing over time.

3	Individual ship operators accident and claims results.

Each of the three possibilities was explored to see if any hard facts could be obtained. Unfortunately, it was concluded that only the third option offered any potentially useful information at the time. The problem with the first two potential sources is that they represent the big picture and we are reduced to averages. ISM implementation does not easily reduce to averages — it is a very specific/individual matter which can only be applied to a particular case. However, even at the individual company level, it is not always possible to obtain meaningful statistics. Captain Eberhard Koch drew attention to the problem within a growing company:

"We are very sure that a certain number of accidents and incidents did not occur since implementation of our SMS. We can just not present any figures. It is of no use trying to read tendencies out of our P&I or H&M damage statistics when simultaneously, over these years, the number and composition of ships under our management has changed"

Each statistical possibility is now explored in some detail so as to describe the problems that presently seem to exist.

Indications from the marine insurance sectors

Ship owners have traditionally split their main insurance requirement into two parts, each part being covered by very different sectors of the insurance industry. The ship itself, with its machinery and other equipment is usually insured on a Hull and Machinery (H&M) policy drawn up on one of the worlds marine insurance markets, such as the Lloyds market in London or with insurance companies. There are also a small number of mutual Hull Clubs. The liabilities, on the other hand, tend to be insured with the mutual Protecting and Indemnity Associations, more usually referred to as P&I Clubs. More than 90% of the deep sea ship owners of the world have their liabilities covered in one of the 13 member clubs of the International Group of P&I Clubs. The owners of cargo being transported by sea tend to arrange insurance cover on one of the insurance markets or through an insurance company.

Accidents or incidents on board ship, of any significance at all, will almost certainly manifest themselves as insurance claims of one description or another. Consequently it would not be unreasonable to expect that insurance claims figures could be examined for the period before and after the Phase One implementation deadline of 1st July 1998 to ascertain whether any detectable trend was developing. Further, if there was a trend then it should be possible to measure the differences. Unfortunately, things were not quite that simple. A number of marine underwriting organisations and companies were contacted, as well as all of the P&I Clubs with a request for statistical data on their claim trends. It became very difficult to obtain meaningful information. The probable reasons for this are discussed presently.

One thing noticeable about the various sectors of the marine insurance industry and their attitude towards the ISM Code — in particular comments on its apparent success or failure — was their almost total silence. The main exception was a series of claims made by the Swedish P&I Club in Goteborg, which have received widespread attention. Indeed the Secretary General of the IMO, Mr. William A. O'Neil, as well as many other journalists and industry leaders, have made extensive reference to the Swedish Club findings as confirmation that ISM can work. The author does not doubt the accuracy or the sincerity of the Swedish Club study. However, he believes that some considerable caution is needed when considering the findings.

In a bulletin published by the Swedish Club in December 2001 it describes the research, involving a comparison of the claim results (on the hull and machinery side) between 274 Phase One and 319 Phase Two vessels, a total of 593 vessels. The number of commercial trading vessels in the world fleet is estimated at about 47,000 (BIMCO/ISF 2000), suggesting that the Swedish Club survey involved approximately 1·3% of the world fleet. The insurers for the other 98·7% of the world fleet were not, presumably, able to confirm similar results.

The Swedish Club explained its own situation as possibly being attributable to the very high quality of tonnage entered in the club which was not representative of the cross section of the world fleet. This may well be the case but the Swedish Club, along with all the other P&I Clubs and H&M underwriters applied very substantial general increases to the calls/premiums during the 2001 and 2002 policy years with more increases to come. Wishing no disrespect to the Swedish Club — because of its relative size to the global market and the effect on its own statistics of just a few more or less major incidents in one year than the next — the results should be viewed with some caution. If the UK P&I Club had made a similar claim

36

then the whole picture might be much more optimistic, purely from a statistical perspective, since the UK Club provides liability insurance to approximately 25% of the world fleet. Since no such claim has been made we are left to draw our own conclusions from the silence.

However, the work of the Swedish Club does need to be taken seriously and appropriate praise given where due. A review of their methodology and results is thus relevant and appropriate here. The Swedish Club provides mutual insurance for both P&I and H&M. The bulletin they published in December 2001 (*Swedish Club Highlights*) focused primarily on the results from the H&M claim trends — although they have been monitoring their P&I results. In the leading paragraph they claim:

"A new study carried out by The Swedish Club confirms that shipowners implementing the International Safety Management (ISM) Code can expect to achieve a reduction in hull claims of 30 per cent, together with similar improvement in the incidence of P&I claims."

In December 1999 the Swedish Club issued the findings of a study comparing claims involving Phase One vessels, which had to comply by 1st July 1998 and Phase Two vessels working towards compliance by 1st July 2002. Their study reviewed claims trends in the three years to 30th June 1999, noting that claims development during the period appeared to be 30 per cent better for Phase One vessels.

The club had predicted that the gap between the claim trends for the two groups of vessels would narrow as the Phase Two compliance deadline approached. Their actual results appear to confirm this prediction. They tabulated the hull claims development since 1995-96 for Phase One vessels compared to Phase Two vessels, as follows:

1995 – 1996 (base year)	100%
1996 – 1997	95·5%
1997 – 1998	85·2%
1998 – 1999	67%
1999 – 2000	70·8%
2000 – 2001	78%

These results are displayed graphically in figure 4.1 overleaf. The Swedish Club observed that in the three years to 30th June 1999, a substantial gap opened between the hull claims incidence rates for Phase One and Two vessels. Within 12 months of the 1998 deadline, Phase One hull claims were running at just 67% of those for Phase

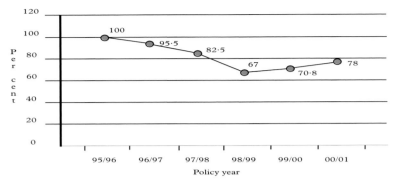

Figure 4.1 — Hull claim development since 1995/96 for Phase One ships in relation to Phase Two ships (1st July 1998 — Phase One deadline)

Two vessels. From that point the gap began to narrow — presumably as an increasing proportion of Phase Two vessels became involved in the ISM Code implementation process. The suggestion is the lowest point of 67% would indicate an improvement of 33% by Phase One vessels compared with Phase Two vessels. If true this would clearly represent significant savings in terms of direct as well as indirect costs of accidents and claims.

They further predict that the gap will continue to narrow as the Phase Two ships complete their implementation process. Following this to its conclusion the two groups should return to 100% coincidence once ISM is fully implemented in both groups.

At a major international conference held in May 2002 Mr. Martin Hernqvist, the Swedish Club's Loss Prevention Officer who had carried out the ISM study, presented a further set of figures which included the P&I results (figure 4.2).

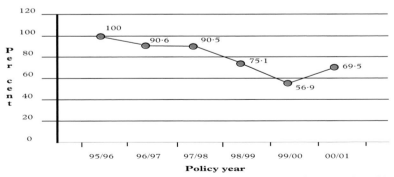

Figure 4.2 — Hull and P&I claims development since 1995/96 for Phase One ships in relation to Phase Two ships (1st July 1998 — Phase One deadline)

38

The Swedish Club also conducted a survey of its membership and although the sample was relatively small — 94 companies replying — it produced some interesting data. The results can be seen in the figures 4.3 and 4.4:

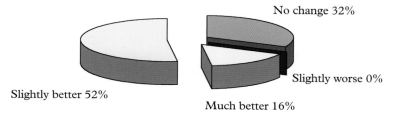

Figure 4.3 — Can you see a change in the incident rate involving your vessels since the implementation of ISM?

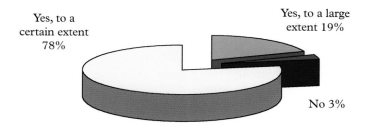

Figure 4.4 — If the incident rate has become better, do you think that has to do with the implementation of ISM?

Three further sets of findings from the Swedish Club survey are worthy of note:

1   The three most important factors for a properly functioning safety management system and improved safety records were:

  •   Commitment from the top management ashore.

  •   Increased safety awareness on board.

  •   Checklists/procedures for key shipboard operations.

2   The reasons for a non-functioning safety management system were:

  •   Too much paperwork/documentation.

  •   People do not know how they are expected to use the system.

  •   People do not believe in ISM.

3   The top four proposals on how to convert a poorly functioning system into a properly functioning one were:

- More ISM training and education.
- Reduce paperwork/documentation.
- Provide seafarers with good examples of ISM in practice.
- Make sure that the accident reporting procedures work and increase feedback from accident reports to the seafarers.

These observations, in many ways, reflect the authors' results and conclusions. There are a number of problems the author has with the Swedish Club claims — although, as stated, not doubting the sincerity with which the club has conducted its research nor the accuracy of the figures used. The original claim that improvements of 30% were being noted between the two groups of ships was made just over one year after the Phase One implementation date. Admittedly, the club had been monitoring trends for three years previously but, even so, it would be unusual for a group of independently operating shipping companies to achieve such implementation success in such a short time. The sample is small and one or two additional major incidents would severely distort the figures.

A factor that cannot be ignored, though it in no way disproves the claim, is that the Swedish Club received much good publicity with the result of their study being widely reported. So, why has no one else made similar claims? The Secretary General of IMO continues to make repeated reference to the study when seeking something to quote to confirm that ISM is working. Repeated references are made to it in newspaper and magazine articles and at conferences and seminars. The fact is that it appears to be the only such claim to have been made. Any of the other P&I clubs would have revelled in such good publicity. Those remaining P&I clubs still cover more than 90% of the worlds' ship owners. If any one of them could have made a similar claim then they almost certainly would. Their silence perhaps tells its own story.

All the other P&I clubs were approached as part of this research. Many responded and indicated that they would provide details of their claims records. Unfortunately, although repeated reminders were sent, only five clubs ended up supplying figures and then it was on the strict understanding that total confidentiality was to be maintained. However, the author is most grateful to those clubs that provided statistics. The review of those figures was extremely useful. The review basically confirmed that there was little by way of measurable trends in the claim figures since Phase One implementation in 1998, although in some instances it might be possible to detect a downward trend in the number of claims but this tended to be accompanied by an upward trend in the value of claims.

Looking for claim trends in this way could also be misleading. Consider, for example, if the membership of a particular P&I was such that all the companies were already operating good safe ships in advance of ISM and didn't generate many claims. In such a situation it would be very difficult to do anything which would nudge the trend downwards; it is more likely to remain steady. The same argument could also apply to individual ship operators.

The reality is that each P&I club comprises many different ship operating organisations. Each organisation progresses at its own pace with ISM implementation and, consequently, the result for the whole club is going to be nothing more than the average across its entire membership. Some ship owners appear to be making good progress with their implementation while many others are taking a little longer. It will take a number of years before collective results can be detected, whether that be across a particular P&I club, or a classification society or a particular flag.

However, the situation is not as simple as that. The whole of the marine insurance industry appears to have been under-funded in recent years i.e. the premiums being paid are not sufficient to cover the cost of the claims and liabilities which have been incurred. This situation has arisen because of two main reasons:

1   Market competition. More underwriting capacity than customers and consequently insurers have been competing for business, even at below break-even prices.

2   The insurance sector, like everyone else, has been badly hit by the very poor return on investment income. As a consequence insurance premiums have been increasing significantly in all sectors during the 2001/2002 policy years.

Almost all the P&I clubs of the International Group of P&I Clubs made a general increase of between 25% and 35%, with some clubs also making additional supplementary calls. One of the leading broking houses, AON, report the following rises in other parts of the marine insurance industry:

- Blue water hull   15 – 33%
- Brown water hull   15 – 25%
- Brown water P&I   15 –30%
- Marine liability   7·5 – 20%
- Excess liability   7·5 – 15%
- Brown water pollution   0 – 15%
- Cargo   10 – 20%
- Cargo storage   10 – 30%

Even if an insurer was in a position to follow the Swedish Club claim and make an announcement that they were seeing a reduction in claims which they attribute to successful ISM implementation, it is suggested that they are unlikely to make any such claim. In the present market climate they might have some difficulty explaining to their members/clients why they then make another announcement that they require a further increase of 25% on all premiums for the second year running!

It proved equally difficult, initially, to obtain meaningful statistics from the H&M insurers. A few years ago the Joint Hull Committee at Lloyds used to produce some excellent casualty statistics. With the retirement of a particular individual, however, those statistics seemed to cease being produced. Part of the problem is that, unlike the P&I sector, H&M insurance is very fragmented. Indeed, the way in which each ship is insured makes statistical data collection from a single reference source very difficult. On each ship there may be 100 or more individual insurers, who may be positioned in different markets. The lead underwriter may take one or two percent and then many others would take smaller lines. In this way each individual insurer limits its exposure and spreads the risks covered over a wide portfolio.

However, a very useful source was discovered in Norway. Central Union of Marine Underwriters in Norway (CEFOR) (www.cefor.no) provide, as a service, the Norwegian Marine Insurance Statistics (NoMIS) whose purpose is to compile and process statistical information. As of 31st December 2001 the NoMIS database claimed to comprise 53,167 vessel years and 20,113 claims. Their website contains a detailed statistical report on those incidents, including international as well as Norwegian business and covers the period 1990—2001. The author is most grateful to CEFOR for their permission to reproduce the graph at figure 4.5 which shows total claims per underwriting year for that period:

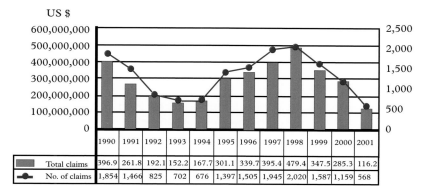

Figure 4.5— Total claims per underwriting year — 1990 to 2001
Source: CEFOR — www.cefor.no

What is very interesting about this particular graph is that it shows very clearly indeed the cyclical nature of marine claims. It is tempting to draw certain inferences and conclusions from the steady downward trend since Phase One implementation in 1998. It is also tempting to suggest that these figures provide a partial explanation for what the Swedish Club had found. Clearly time will tell whether the recent downward trend is just part of the cyclical pattern of marine insurance claims or whether it really is heralding good news.

A number of respondents were quite cynical about the marine insurance industry and seemed to infer that there was some sort of conspiracy taking place. For example an interesting observation was received from a manager of a Greek shipping company who said:

"As far as insurance companies are concerned, ISM has achieved its goals by 100% as they now manage to pay less than what they did in the past ..."

A similar perception was put forward by an Indian operations manager in a shipping company who said:

"At the risk of sounding hopeless, I honestly believe that the one thing that the ISM Code has surely achieved is creating an hitherto absent paper trail that helps pinpoint blame when accidents occur, helps insurance companies shy away from paying on claims and aggravates the already stressed-out lives of modern day seafarers ..."

The author received other, similar, reports but did not receive details of any actual or specific incident where insurers had declined to pay claims on account of ISM violations. Presumably the reports were based on personal experience but clearly run quite contrary to what the ISM Code is about and what it is trying to achieve.

43

Certainly the marine insurance industry will be delighted if the ISM Code is a success, because that will be beneficial to both the shipping and the insurance industries.

Indications from Port State control MOUs

IMO promulgate general guidance to Port State control by resolution A.787(19) but in December 1998 the IMO produced a set of 'Interim Guidelines for Port State Control Related to the ISM Code' by way of circular MSC/Circ.890 / MEPC/Circ.354. Numbered paragraph 3 of those guidelines states:

"3. Port States should recognize that Port State control related to the ISM Code should be an inspection and not an audit. The ISM Code has been developed to promote a safety culture and is not intended to penalise those ships/operators whose safety management systems embrace the principles and requirements of the ISM Code"

The intention is that the Port State control should confirm that there is a SMS in place and which appears to be working.

A Concentrated Inspection Campaign (CIC) was carried out by the Paris MOU, in conjunction with the Tokyo MOU, following Phase One implementation on 1st July 1998. The data from that CIC remains an important source, specifically linking a detention to a non-compliance with the ISM Code. The campaign ran from 1st July to 30th September 1998. A further CIC was conducted following the final deadline for Phase Two implementation on 1st July 2002. At the time of completing this manuscript the data had not been made available.

The first results of the 1998 CIC showed that a total of 1,575 eligible ships were inspected during the campaign. A uniform questionnaire had been used by PSCOs to test key elements of the ship's safety management system. A total of 81 ships were detained in port for major non-compliances in their system. Three ships were banned from the Paris MOU region for not having ISM certificates on board and a safety management system in place. These ships would not be allowed to enter any of the Paris MOU ports until evidence was provided that a certified management system was in place.

Bulk carriers were the largest category of ship found not to comply with 58 being detained out of 722 inspected. Chemical tankers, oil tankers, passenger ships and gas tankers figured to lesser degrees. Twelve Flag States, with more than 10 ISM applicable inspections, showed the following rates of detentions:

Turkey	16.4	St. Vincent & Grenadines	12.1
Russia	10.3	Bahamas	7.4
Cyprus	6.7	Panama	6.5
Philippines	5.6	Liberia	4.8
Malta	4.3	Greece	4.1
Italy	3.3	Norway	1.0

Since certification for the ISM Code provisions had been carried out to a large extent by classification societies on behalf of Flag States, the results of the Paris MOU analysis indicated the detention rate by class in relation to a minimum number of 10 ISM applicable inspections as follows:

Bureau Veritas	11.3	Hellenic Register of Shipping	10.5
China Classification Society	9.5	Russian Maritime Register of Shipping	9.4
Lloyds Register of Shipping	9.3	Nippon Kaiji Kyokai	7.4
Polski Rejestr	6.7	American Bureau of Shipping	5.7
Registro Italiano Navale	4.0	Det Norske Veritas	3.4
Germanischer Lloyd	1.2		

In the inspected areas of the management system on board, the Paris MOU analysis showed that the most frequent major non-conformities were:

- Certificates and particulars not in order. 21%
- No certificates on board. 17%
- Senior officers not able to identify designated person. 16%
- No maintenance routine and records available. 12%
- Senior officers not able to identify company responsible for the operation of the ship. 9%
- Programmes for drills and exercises to prepare for emergency actions not available. 8%

The campaign was too soon after Phase One implementation to draw firm conclusions about the extent of non-compliances. The initial impression, however, provided good support for the strong stance on enforcement by the Paris MOU and others, to show that it had been effective in driving away non-compliant ships — or at least those ships which had not gone through a certification process. Prior to the 1st July 1998 deadline there had been predictions that there would be serious congestion of the ports with fleets of banned, non-compliant, ships. That did not materialise.

The 2001 Paris MOU Annual Report recorded 18,681 inspections carried out on 11,658 ships. On ISM it stated that there had been a steep increase in safety management defects which, over three years, had increased by 150%. Clearly this is an alarming figure and is depicted in figure 4.6, derived from the 2001 Paris MOU Blue Book.

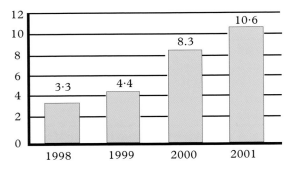

Figure 4.6 — Ratio of deficiencies to individual ships x 100

A regular criticism of the Port State control system is its lack of consistency and uniformity in its interpretation of what it is supposed to be doing. Clearly resolution A.787(19) and MSC/Circ.890 / MEPC/ Circ.354 were an attempt to address this issue and the secretariat of the Paris MOU very kindly provided the author with a copy of a document – 'Port State Control Instructions – PSCC34/2001/01 dated 11 May 2001 – Guidelines for the Port State Control Officer on the ISM Code'. This is an extremely useful document describing how the PSCO should approach an inspection when ISM factors are under consideration. An extract from that document is set out below:

A. Initial inspection
1. During all routine PSC inspections, a check should be made that the ship has ISM certification on board, in accordance with the ISM Code. The PSCO will at the initial inspection examine the copy of the Document of Compliance (DOC) issued for the Company, and the Safety Management Certificate (SMC) issued for the ship. A SMC is not valid unless the operating company holds a valid DOC for that ship type. The PSCO should also check that required audits and endorsements have been made to the certificates.
2. The PSCO should in particular verify that the type of the ship as

reflected in the SMC is included in the DOC, and that the company's particulars are the same on both the DOC and the SMC.

3. If a vessel has been issued with interim certificates the PSCO should check whether it has been issued in accordance with the provisions of Section 14 of the Code. Though the safety management system may not meet the items 4 – 17 listed in Part 2, a documented system should be in place and the PSCO should use his professional judgement in deciding whether a more detailed inspection is necessary.

B. More detailed inspection

A more detailed inspection of the SMS shall be carried out when clear grounds are established. Clear grounds include absent or inaccurate ISM certification or detainable deficiencies in other areas. Many non-detainable deficiencies may also be evidence of a deficient management system and the PSCO should use his professional judgement in deciding whether these warrant a more detailed inspection. When carrying out a more detailed inspection the PSCO may utilise the following to verify compliance with the ISM Code.

The following questions are not a checklist but contain examples of areas which could be inspected by the PSCO. (Each question has an explanatory note accompanying the guidelines)

1. Is the ISM Code applicable to the ship?

2. Is ISM certification on board?

3. Are certificates and particulars in order?

4. Is there a Company safety and environmental protection policy and are the appropriate personnel familiar with it?

   Ref: Section 2.2 of the Code

5. Is safety management documentation (e.g. manual) readily available on board?

   Ref: Section 1.4 of the ISM Code

6. Is relevant documentation on the SMS in a working language or language understood by the ships personnel?

   Ref: Section 6.6 of the ISM Code

7. Can senior officers identify the Company responsible for the operation of the ship and does this correspond with the entity specified on the ISM certificates?

47

Ref: Section 3 of the ISM Code

8. Can senior officers identify the designated person?

Ref: Section 4 of the ISM Code

9. Are procedures in place for establishing and maintaining contact with shore management in an emergency?

Ref: Section 8.3 of the ISM Code

10. Are programmes for drills and exercises to prepare for emergency actions available on board?

Ref: Section 8.2 of the ISM Code

11. How have new crew members been made familiar with their duties if they have recently joined the ship and are instructions which are essential prior to sailing available?

Ref: Section 6.3 of the ISM Code

12. Can the master provide documented proof of his responsibilities and authority, which must include overriding authority?

Ref: Section 5 of the ISM Code

13. Does the ship have a maintenance routine and are records available?

Ref: Section 10.2 of the ISM Code

14. Have non-conformities, accidents and hazardous situations been reported to the Company and has timely corrective action been taken by the Company?

Ref: Section 9.1, 9.2 of the ISM Code

15. Are procedures in place for maintaining the relevant documentation?

Ref: Section 11 of the ISM Code

16. Are procedures in place for internal audits and have these been carried out?

Ref: Section 12 of the ISM Code

PSCOs should not normally scrutinise the contents of any non-conformity notes resulting from internal audits.

17. Do detainable deficiencies or many deficiencies, if found, indicate a failure of the safety management system?

C. Follow-up action

1. No ISM certification on board

In Annex 1 a flow chart is presented which shows all necessary steps after an initial inspection when the COC and/or SMC are missing. The chart includes the requirements of the Council Directive 95/21/EC.

2. No valid ISM certificate on board

When the ship's ISM certificates are invalid, e.g. periodic verification has not taken place or discrepancies exist between the DOC and the SMC, the Flag State and the Company will be requested to take appropriate rectifying action. The principles outlined in Section 9 of Annex 1 to the MOU with regard to detention and rectification of deficiencies are applicable.

3. Detainable deficiencies in hardware and/or operational areas

3.1   The normal procedure in accordance with section 9 of Annex 1 to the MOU will be applicable.

3.2   Detainable deficiencies and multiple non detainable deficiencies may indicate a failure of the safety management system. However, the PSCO cannot automatically conclude that the system has failed. The PSCO should examine the relevant areas of the system to identify non-conformities.

3.3   Non-conformities shall be recorded on the PSC inspection form as indicated in Part 5 (SIRENAC Codes). The PSCO shall ask the Flag State to issue and follow up non-conformity notes. Issuing classification society informed if appropriate.

3.4   Prior to detention being lifted, the Company shall report to the Port State authority which corrective action will be taken regarding the non-conformities which have been reported.

3.5   Non-conformities have to be rectified within three months.

3.6   Major non-conformities have to be rectified before sailing.

D. Areas which may warrant detention.

The following items may be considered as major non-conformities [footnote 2] and would make the vessel liable for detention. This list is not considered exhaustive but is intended to give an example of relevant items.

Section of the ISM Code:

13 ISM certificates not on board.

13 Company on the DOC not the same as on the SMC.

1.4 Safety management documentation not on board.

6.6 Relevant safety information not in a working language or a language understood by the crew.

3 + 4 Senior officers unable to identify operator and designated person (ship/shore system breaks down with this).

8.3 No procedures to contact the Company in emergency situations.

8.2 Drills have not been carried out according to program.

6.3 New crew members are not familiar with their duties within the SMS.

5 Master's overriding authority not documented and master unaware of his authority.

10.2 No records of maintenance kept or no evidence of maintenance being carried out as indicated in the records.

The reason all this has been set out in detail is because it is important to understand what PSCOs are recording as ISM deficiencies. Certainly all of the above would be considered ISM deficiencies but consider a situation whereby the PSCO identified deficiencies with the hardware, maybe lifeboat davits that had seized, for example or fire extinguishers that were empty. These are certainly deficiencies under the life saving appliance rules and perhaps other sections of SOLAS and would no doubt be recorded as such in the defect notice. However, they would also clearly point to a seriously defective safety management system. If the SMS had been functioning then the life saving appliances would never have been allowed to fall into such a state of disrepair. However, such deficiencies are unlikely to be recorded as ISM deficiencies. Consequently, in the author's view, the PSC statistics may not necessarily be an accurate indicator of ISM related problems.

There is another concern with regard to the PSC records of ISM deficiencies. It is not at all surprising that the number of recorded ISM deficiencies is increasing with the passage of time, rather than decreasing as might have been hoped or expected. The most likely reason for the increase is not that there are actually more deficiencies but rather that more inspections are taking place on more ships. In addition the PSCOs themselves are becoming much more sophisticated, knowledgeable and skilful in inspecting as well as auditing and interrogating management systems. Indeed a number of seafarers were almost boasting in their survey questionnaires as to how they were able to hide problems in the SMS from the PSCO. It can be anticipated that they will continue to develop their sophistication and start identifying an increasing range of ISM deficiencies. Many have already understood the point being made above with regard to defects in the hardware being indicative of problems with the software — i.e. the SMS. For these reasons the author has doubts about placing too much reliance on the PSC detention figures at this time as being any sort of indicator relating to ISM compliance, although more detailed access to the actual detention documents might provide the necessary clarification.

Indications from individual ship operators

During the course of his research the author had the privilege of encountering a relatively small number of ship operating companies where the SMS really had become well established and the rewards were coming in. These companies were not the oil majors or the old liner operators. They were relatively modest organisations, quite typically operating a dozen medium sized ships, who had made 100% commitment to ISM. Having said that, even old established companies such as A.P. Moller conceded that whilst they considered they had managed safety perfectly well for many years, through ISM implementation they had managed to scrap 25% of their former circulars and circular letters.

Details received from other companies were perhaps a little briefer but still of interest — such as the comments of an Indian manager in a ship operating company:

"...As far as our Company is concerned, we have gone a long way towards the effective implementation of ISM Code ... I feel that compared to the past, presently our operations are more organised. Our operating cost has reduced; there are practically no time loss accidents and there is a more healthy atmosphere to work both on board and ashore ..."

A similar experience was reported by a Swiss manager:

"...already prior implementation of ISM aboard our vessels, the safety culture was an important matter, it is true that after ISM implementation, accident aboard have been greatly reduced, maintenance boosted up and mechanical damage to property improved ... I definitely confirm that this system aboard our vessels is working but need continuous control ...

The level of achievement and improvement which might be possible is clearly going to vary enormously from company to company and from ship to ship. It all depends where the starting point might be. Those who will have most to gain from a properly implemented SMS will be companies previously operating at a relatively low level of safety management. However, even those companies who believed they were already managing safety at a high level have found that a whole range of improvements have been possible, which raises their own level of safety even higher. An operations manager in a shipping company shared his reflections on this very issue:

"ISM has different meaning for different people. We use ISM in an enhanced form to manage our entire operation, i.e. it is the way we work and relates to all aspects of operations, not only safety and the protection of the environment. For us it is successful. Many companies, however, require only the certification and do not actively use their SMS to enhance their management. This attitude will change with time and therefore in the longer view the ISM Code will contribute to a better managed and more professional ship management industry. Many operate at a higher level already but ISM will drag the base level higher. In summary we are better off with ISM than without it."

There were companies clearly trying to move forward but encountering obstacles such as reported by the following ship manager:

". . . after five years of ISM I feel our system is growing top heavy, mainly due to complying with requests from auditors — we are about to conduct a major review to stop the tail wagging the dog . . ."

It is very important always to remember that the SMS is intended to be a dynamic process which is constantly evolving. It should be ready and capable of change should change be needed to improve the efficiency of the system. Others reported situations that were not so encouraging and in some cases reflected almost a sense of hopelessness — such as this Cypriot ship operator:

". . .unfortunately it has created a heavy burden of paperwork to all people ashore and the captains aboard the vessels (specially the small ships with limited crew) without significant change to the safety and environmental protection . . ."

It would appear that many ship operators and seafarers have found themselves in very similar dilemmas. The conclusion often reached is that the ISM Code is flawed and that all the paperwork and inefficiencies associated with it are the fault of the Code. As explained elsewhere, the Code does not call for vast amounts of paperwork to be generated. These problems are often associated with a poorly structured and badly implemented SMS, particularly those which are completely alien to the particular Company which had been bought off the shelf. Ship operators and seafarers facing such problems should take a long hard look at their SMS. If they conclude that the SMS they have is a good and efficient system but the staff they have still cannot manage the system adequately then they probably do have a serious manning problem. That problem is probably acute and in need of urgent corrective action before it leads to a serious accident. The response may well be that the ship operator believes that he cannot afford to employ any more staff. The author sympathises with that dilemma but would suggest that the cost will be very small compared with the alternative. In this regard the case studies in Chapter 5 should be read carefully — particularly the *Eurasian Dream*. In addition to losing the right to rely on any of the Hague Visby defences on account of having an ineffective SMS, ship operators may find that insurers start using similar tests and refuse to pay claims in appropriate cases.

Chapter 5

THE ISM CODE AND RECENT COURT DECISIONS

Introduction

To bring alive the potential impact of some issues arising out of ISM implementation, in a practical way, consider three recent court cases. In particular, these cases address:

- ISM as the benchmark against which operational good practice will be measured.

- Evidence and audit trails.

- The test of reasonableness.

The three cases considered are:

1   The *Eurasian Dream* — QBD (Com. Ct.) (Cresswell J) — 7 February 2002.

2   The *Torepo* — QBD (Admiralty Ct.) (Steel J) — 18 July 2002.

3   The *Patraikos II* — Singapore High Court — 9 May 2002.

All three cases were cargo related liability incidents where the issue was the question of seaworthiness and whether the respective ship operators had exercised due diligence to make their ships seaworthy within the terms of the Hague Visby Rules, the relevant section of which, Article III Rules 1 and 2, reads:

---

1.   The carrier shall be bound, before and at the beginning of the voyage, to exercise due diligence to –

a)   Make the ship seaworthy.

b)   Properly man, equip and supply the ship.

c)   Make the holds, refrigerating and cool chambers, and all other parts of the ship in which goods are carried, fit and safe for their reception, carriage and preservation.

2.   Subject to the provisions of Article IV, the carrier shall properly and carefully load, handle, stow, carry, keep, care for, and discharge the goods carried.

The initial burden will be upon the cargo claimant to establish a prima facie case of unseaworthiness and to demonstrate that they have suffered a loss as a consequence of that unseaworthiness. Article IV Rule 1 of the Hague Visby Rules provides that the carrier shall not be liable for loss or damage arising from unseaworthiness provided they can demonstrate that they exercised due diligence to make the vessel seaworthy. What this means in plain English is that if the ship owner did all that they reasonably could to make sure that they provided a ship and crew that should have been able to deliver the cargo at destination in the same condition as it was in when loaded, then they will be entitled to rely upon certain exclusions and defences. There is a list of 17 exceptions set out in Article IV Rule 2 — part of which is set out below with the exceptions most relevant to the case studies:

> 2. Neither the carrier nor the ship shall be responsible for loss or damage arising or resulting from —
>
> (a) Act, neglect, or default of the master, mariner, pilot, or servants of the carrier in the navigation or the management of the ship.
>
> (b) Fire, unless caused by the actual fault or privity of the carrier.
>
> (c) Perils, dangers and accidents of the sea or other navigable waters . . .
>
> (q) Any other cause arising without the actual fault or neglect of the carrier, or without the fault or neglect of the agents or servants of the carrier, but the burden of proof shall be on the person claiming the benefit of this exception to show that neither the actual fault or privity of the carrier nor the fault or neglect of the agents or servants of the carrier contributed to the loss or damage.

The author believes that all three cases have direct relevance to ISM and provide valuable guidance from the courts on the way in which they will examine cases in light of ISM. However, it should be pointed out that the ISM Code was not mandatory on the *Eurasian Dream* or the *Torepo* when the incidents actually occurred. It would appear from reading the law reports that the respective judges clearly had ISM principles in mind when considering the cases and when arriving at their judicial decisions. In the *Eurasian Dream* judgement it is said that the ISM Code is " . . .a framework upon which good practices should be hung . ." and " . .a prudent manager/master could very well organise their companies vessels work following those guidelines . ." (*Eurasian Dream* judgement paragraph 143).

The factual information in the case studies below is based entirely on the respective law reports. Clearly, by their very nature, such reports must be selective in what factual information they include. It is always possible therefore that there may be some slight errors in fine detail

— for which the author extends his apologies. However, the author believes that the main principles, which are the important features, are accurately represented. It is necessary to appreciate, however, that these case studies are not intended to represent a full review of any of the incidents. Anyone wishing to obtain full details of the judgements are referred to the relevant law report.

Case Study 1 – The *Eurasian Dream*

Brief facts:
- The *Eurasian Dream* was a pure car carrier, of the large box-shape design.
- She was discharging a cargo of new motor vehicles in Sharjah in the United Arab Emirates.
- A fire started in car deck number 4.
- The crew were unable to contain the fire.
- The ship was abandoned and towed off the berth.
- The ship and all the cargo were a total loss.
- There were no serious injuries to personnel.

The basis of the claim:
- The cargo owners/subrogated underwriters claimed for damage to their cargo and other losses which they suffered as a consequence of the fire.
- The basic argument was that the ship was unseaworthy and that the carrier had failed to exercise due diligence to make the vessel seaworthy.
- Specifically, they argued that the ship was unseaworthy on many counts, including:
  — Vessel's equipment.
  — Competence of master and crew.
  — Adequacy of documentation.

Summary of events:
Before undertaking an evaluation of the evidence it is worth reviewing briefly what happened and what went wrong. By doing so the significance of the judge's criticism should make more sense.

During the relevant period the vessel was discharging cars. The third officer (3/O) was the duty cargo officer and he had an able seaman (AB) on duty with him. Both the 3/O's and AB's stories changed, on a number of occasions, as to exactly where each was at the time when the fire started. Some versions said the 3/O was on the car deck and witnessed the fire start whilst other versions had him in the mess-room taking a snack. The 3/0 did have a walkie-talkie radio but the other three sets on board were being used by the engineers who were involved in a bunkering operation. The 3/O had no means of contacting anyone else on his radio.

The cars were carried with the minimum amount of fuel in their tanks and with their batteries disconnected. In order to speed up the discharging operation the shore stevedores had adopted a very bad and unsafe practice of pouring a small quantity of petrol into the carburettor of the vehicle and then jump starting the car from a powerful battery on the back of a pick-up truck. Invariably fuel gets spilt in such operations and with the sparks being generated from the battery leads there is a very high risk of a fire. Clearly such practices should be totally prohibited. There was no evidence that any attempts were made to stop the stevedores from carrying out this very bad practice.

It was not proved conclusively, but the conclusion reached by the fire experts was that the fire probably started when some spilt fuel ignited. If a working fire extinguisher had been available then the fire should have been capable of being extinguished fairly easily and quickly. There were a number of partially discharged fire extinguishers found near the seat of the fire by the fire investigators. They concluded that the extinguishers had probably not worked properly.

It took a significant length of time before the alarm was raised and then there was much confusion as to where the fire was or what to do about it. By this time the fire was quickly spreading. The master ordered fire hoses to be run out — this proved to be ineffective. Breathing apparatus sets eventually arrived on the scene but these were too few in number and probably defective. It transpired that no serious or informed emergency practice drills had been carried out on board — particularly on the most basic of all potential emergencies — how to fight a fire. Eventually, the master decided that the fire was out of control and decided to get everyone ashore. Almost as a parting gesture he ordered the CO^2 bottles to be discharged — since the very large loading door was open this smothering gas would have quickly escaped and would have been of no use. Local tugs pulled the burning vessel off the berth and allowed her to burn out in the relative safety of the outer harbour.

The method of fighting a fire on these types of vessels is very different from the method in conventional cargo ships or bulk carriers. The ship is divided up into car decks vertically, with a number of transverse bulkheads. In these bulkheads are gas tight doors such that, in the event of a fire, they can be closed across adjacent bulkheads to completely seal off small sections. Of course, any personnel in the compartment must be evacuated prior to closing the doors. Once the section is sealed CO_2 is injected and any fire would very quickly be brought under control.

There do not appear to have been any instructions on board on this procedure or how to operate the CO_2 smothering or the gas tight doors. It does not appear that the master, chief engineer or anyone else knew about the procedure or how to operate the CO_2 system or the gas tight doors.

Evaluation of the evidence:

- *The witnesses:*

In a court case such as this there are two main types of evidence to be presented — written/documentary evidence and witnesses. The witnesses are either witnesses of fact, i.e. people who were actually involved in or around the incident, or expert witnesses — people with specialised knowledge or skills who can provide the benefit of their superior knowledge for the court to take into account when considering the evidence. The impression that a witness makes on the judge can be very important indeed. If the witness starts changing his story during cross examination, or for other reasons, the judge starts to consider that the witness is not telling the truth. This will, inevitably, be very prejudicial to one side of the case.

Witnesses can be put under considerable pressure during cross-examination. Barristers/advocates are skilled in extracting the truth from witnesses. Consequently, the only way to be sure of successfully withstanding interrogation or cross examination in the witness box is to be completely truthful and stick to the truth and the facts. Witnesses must avoid being drawn down the road of subjective speculation, however tempting that might be.

In this case the master, as well as a number of officers and crew from the ship and the Designated Person from the office ashore, were called as witnesses of fact. A lecturer from a maritime academy, who taught fire fighting as one of his subjects, appeared as an expert witness appointed by the carrier and a car carrier expert was appointed as expert witness on behalf of the cargo claimants. In his judgements the judge had the following comments to make about the different witnesses who had been presented from the ship operators side:

- The master — unsatisfactory.
- Third officer — profoundly unsatisfactory.
- Chief engineer — profoundly unsatisfactory.
- Able seaman — profoundly unsatisfactory.
- Designated person ashore — most unsatisfactory.
- Expert witness — did not have relevant expertise.

During the cross examination of the witnesses, a number of them started to change their original stories significantly, to such an extent that the judge was left wondering what he could or could not believe. In addition, it started to become apparent that deliberate attempts had been made to tamper with important pieces of evidence, including an instruction from the ship manager's office to the chief engineer to return on board after the fire had been extinguished to adjust the position of certain valves before the fire investigator got onboard. Once a judge starts disbelieving witnesses it is then very difficult to restore sufficient confidence to resume a position of trust.

- *Vessels equipment*

During the course of the investigation and court hearing it became apparent that a number of important fire fighting appliances probably failed and other pieces of safety equipment proved to be inadequate. One of the major problems for the ship operator in defending the claim was that there were no records available of when fire extinguishers, for example, were examined, tested or recharged. This applied to a number of other items of equipment such as gas tight doors and CO_2 smothering equipment. The expert witness put forward by the claimants was able to convince the judge that there were significantly fewer walkie-talkies on board than was actually needed. The same applied to breathing apparatus sets. There was no evidence of any risk assessments ever having been carried out which might have picked up those deficiencies.

- *Competence of master and crew*

The master's certificate of competency does not appear ever to have been called into question. However, his ability and qualifications to command a car carrier certainly were. The judge found the recruiting practices of this company incredible, to say the least. Apparently the ship managers needed a master for *Eurasian Dream* and sent word to their manning agents in Manila to find one. It did not seem to concern anyone at the time, but the master they found had not sailed on *Eurasian Dream* before, or any sister ships. Indeed, he had neither sailed on car carriers before nor with that company before. Apart from the manning agent, he had not met anyone from the Company but was just instructed to go and join the vessel. The only instruction he received from the Company was a letter, basically telling him to read all the manuals when he got on board.

It transpired that there were well over a hundred manuals that he would have had to read. It was estimated that if he did nothing else apart from read manuals then he would need two to three weeks to complete the task. He was apparently given no specific instructions about safety procedures with regard to loading, carrying or discharging motor vehicles. He was given no instructions on how the CO_2 system worked and, apparently, wasn't even told of the existence of the gas tight doors, their purpose or how they worked. The ship operators were unable to produce evidence to describe any adequate procedures they might have had in place to provide relevant familiarisation or training. There was no evidence that the crew had ever undergone any meaningful emergency drills or exercises such as fire fighting. A detailed quote from the judgement is set out in the subsection on the adequacy of documentation below.

It also became apparent during the examination of the witnesses and the documentation that none of the other officers or crew members had what could be described as an adequate knowledge or understanding of safe cargo operations on this type of vessel. The judge concluded that the master and crew were ignorant of:

- Fire hazards in car carriers.
- The need to supervise stevedores.
- Procedures for dead car operations.
- CO_2 procedures.
- Gas tight door operation.
- *Adequacy of documentation.*

As stated earlier, the ISM Code was not mandatory for this type of vessel at the time of the incident. Car carriers are ISM Phase Two ships and, consequently, would have until the 1st July 2002 to comply. The incident occurred on 23rd July 1998.

However, the particular ship managers operating *Eurasian Dream* also operated Phase One ISM ships and had produced procedures manuals for those ships. It appeared that these were generic manuals, apparently produced with the intention that they would apply to all their vessels. It appeared that copies were put on board all their ships.

In his judgement, Justice Cresswell discusses this documentation in some detail. Although rather a long quote, it is felt worthwhile including in order that the full significance of what he was saying can fully be appreciated:

"151(12) – (18) It was of fundamental importance that the vessel be provided with a ship specific manual dealing with fire prevention and control. No such manual was provided to the Eurasian Dream.

The vessel was provided with a large amount of irrelevant and/or obsolete documentation. Such documentation related (for example) to vessels other than car carriers. Such documentation was potentially misleading. For efficiency and competence of response, only one code or set of procedures should have been prescribed for the master of a pure car carrier.

The documentation placed on board by Univan was too voluminous to be digestible. The master was directed by a standard form briefing letter to read all the literature on board the vessel. This was an inadequate means of instructing the master for the following reasons:-

(a)   It was not given to the master in advance of his attendance upon the vessel.

(b)   It did not cater for the special position of Captain Villondo, who had no prior experience of car carriage, car carriers, the Eurasian Dream or Univan.

(c)   The direction of the briefing letter required the master to read a vast amount of documentation, including Univan manuals which ran to hundreds of pages and about 100 technical equipment manuals.

(d)   The task of reading the Univan manuals would have occupied two to three weeks of the master's time whilst on board the vessel.

(e)   the briefing letter ought either to have summarised all the key guidance to be given to the master in relation to emergency procedures or to have directed him in a focused manner to the relevant manuals or parts of manuals dealing with such matters.

The emergency procedures manual (and the other Univan manuals) failed to give guidance as to:

(a)   The supervision of stevedores.

(b)   The importance of gastight doors as fire-fighting equipment.

(c)   The efficient use of the CO_2 system (including the speed with which it should be deployed and the steps to be taken to permit such deployment).

(d)   The evacuation of personnel.

Instead, the manual contained guidance for fighting fire on other types of vessels. Such material was irrelevant and the manual was put to one side by the master on this basis. However, if acted upon, the manual was misleading and dangerous: it stated (for example) that, in the laden condition, there was little that the emergency response team could do in the event of a fire and made no mention of any steps which might be taken to fight such a fire.

In accordance with SOLAS, fire-fighting instructions and procedures in particular should have been concentrated in one concise and clear manual, catering specifically for the Eurasian Dream. The master himself complained of the fact that he had not been given such a manual. Univan should have provided the vessel with clear checklists of the essential actions to be taken in the event of fire: (a) at sea (b) in port.

The vessel was not, but ought to have been, provided with specific documentation dealing with:-

(a)   The characteristics of car carriers in general and the Eurasian Dream in particular.

(b)   The carriage of vehicles in general and on the Eurasian Dream in particular.

(c)   The danger of fire on car carriers.

(d)   The precautions to be taken to avoid fire on car carriers, including:

(i) Instructions for the safe handling of second-hand vehicles.

(ii) Instructions for the supervision of stevedores and the prohibition of hazardous activities by stevedores or others, such as simultaneous and proximate jump-starting and refuelling operations in the same area or on the same vehicle.

(e)   The importance of gastight doors in fire fighting.

(f)   The importance of using CO^2 as a front line defence and without delay in the event of a deck fire and simple instructions for its use.

(g)   Procedures for evacuating the fire zones or keeping personnel out of such zones."

A reasonably prudent owner, knowing the relevant facts, would not have allowed the *Eurasian Dream* to put to sea with the master and crew in their state of knowledge, training and instruction.

Even though the vessel did not have to comply with the ISM Code, the judge appears to have adopted the position that since the procedure manuals had been put on board then there must have been an intention on the part of the ship operators to use them as the basis of setting up a SMS. Indeed at paragraph 151(12) the judge actually says that a fire fighting manual (which was not on board) should have been. The judge also seems to have adopted the view that the ISM Code had already set the standard as to the minimum level of ship operation and therefore adopted the general principles as a sort of benchmark. To what extent that is the case or whether it is speculation is somewhat academic now since the Phase Two deadline has passed. In any event there can be little doubt, following a reading of the judgement, that the judge had the ISM Code very much in mind when he considered the evidence and the operational practices in this case.

The message which the judge clearly sent out is that those ship operators who dump off-the-shelf, stereotype manuals on their ships or otherwise pay only lip-service to ISM had better look out. The courts are clearly alert to the intentions and philosophy of ISM as well as what it should be achieving in practice. Inadequate and inappropriate manuals may very well render the vessel unseaworthy. In paragraph 127 of the judgement, Justice Cresswell said:

" Seaworthiness must be judged by the standards and practices of the industry at the relevant time, at least so long as those standards are practical and reasonable".

The standards and practices of the industry in this respect are now the ISM Code. It is also clear from this judgement that the managers have a responsibility to ensure that the master and crew are properly familiar with the ship and the cargo to be carried. In addition, the managers do not discharge their responsibility simply by providing large volumes of documents for the master to read.

The judge was very critical of the ship operator as well as the staff, both ashore and on board the ship. The judge concluded that the vessel was unseaworthy on account of an accumulation of serious problems to do with:

- The vessel's equipment.
- The competence of the master and crew .
- The adequacy of the documentation.

On that basis the judge allowed the cargo owners to succeed with their claim.

Case Study 2 — *The Torepo*

Brief facts:
- *Torepo* was a product tanker.
- On a loaded voyage from Argentina to Ecuador via the Patagonian Channel.

63

- Vessel grounded.
- Vessel had to be salvaged.
- There were no serious personal injuries.
- There was no pollution or loss of cargo.
- The cargo owners attempted to recover their contribution to the salvage.

The basis of the claim
- The cargo owners/subrogated underwriters were claiming for a recovery of their contribution towards the general average/salvage expenses incurred in the refloating operation.
- The basic argument was that there had been a breach of the contract of carriage, in that the ship was unseaworthy and the ship operators had failed to exercised due diligence to make the vessel seaworthy.
- Specifically, they argued that there were many factors which contributed towards that unseaworthiness, including:
    - No proper bridge team management.
    - No proper system for instructing crew in navigating in confined waters.
    - No proper passage plan for that part of the voyage during which the incident occurred.
    - The vessel was not equipped with adequate charts.
    - The echo sounder was defective.

If the cargo claimants were successful with their arguments they would be entitled to recover their contribution to the general average and salvage expenses.

The case was heard before the Admiralty Judge — the Honourable Mr. Justice Steel — a very experienced lawyer and judge in shipping related matters. It is interesting to note that, in his judgement, he was clearly very critical of the way in which the claimants had set out their claim and their allegations. On the whole they were unspecific and unsubstantiated.

The judge was also very critical of the expert witness that the claimants produced to comment specifically on the navigation systems and practices on board *Torepo*. Basically the judge considered the navigation expert to lack realism. This is actually a very interesting, important and enlightened observation on the part of the judge. The expert witness put forward a case for almost absolute perfection. Justice Steel made it clear that neither he, nor the court, nor ISM expects perfection, but best practice will be expected. It comes back to the old problem of reasonableness and what might be reasonable within the context of this particular case. Before we venture off to evaluate the evidence, let us look a little at the background to this incident which happened in July 1997.

Summary of events:

Torepo was a product tanker more than 20 years old of about 25,000 tons deadweight. She was apparently, in relative terms — bearing in mind her age — a well maintained and well run ship. This was confirmed by over 50 vetting inspections by oil majors to which the ship had been subjected during the three years prior to this incident.

On this particular occasion the vessel was close to Buenos Aires in the Argentine and was to load a full cargo of gasoline for Ecuador. It was recognised by the master on board and his navigator, the second mate, that there were at least three routes the ship could take. The first option was to go north and through the Panama Canal, which would certainly provide the best weather option at that time of year (midwinter in the southern hemisphere). The second option was to sail far south and go round Cape Horn but this would almost certainly involve bad weather. The third option was to cut across the tip of Chile and go through the Patagonian Channel. Up until the time of sailing the charterers would not make a decision as to which route they wanted the vessel to take. The second mate prepared the passage plans for each of the three possibilities. However, he realised that they did not have on board all the appropriate navigation charts. In particular, the British Admiralty charts that were being used on board did not adequately cover the Patagonian Channel. The charts required were local Chilean charts. The master attempted to order these from the local chart agent in Buenos Aires but without success. The ship managers tried to obtain them in Europe and the master tried through the local British Admiralty chart agent in Montevideo, Uruguay. However, all these efforts were in vain as it transpired that these charts were only available inside Chile.

Torepo loaded the cargo and set sail, heading south. Eventually the charterers confirmed that the vessel was indeed to proceed through the Patagonian Channel. Local agents were appointed and the master ordered local pilots and also asked that those pilots bring with them copies of the relevant charts.

Evaluation of the evidence:

- The witnesses

  The main witness of fact was the master of the *Torepo*. The master had served with the Company for six years and had sailed previously on *Torepo* as chief officer. The voyage in question was the master's first trip in command. It would be quite normal in such circumstances for the master to try his very best to make sure that everything ran smoothly during the voyage. He would expect people, both in the office ashore and on board ship, to be watching him very closely and he would want to demonstrate that the decision to promote him to master was a correct decision. The judge found the master to be "an intelligent and capable man who responded to cross examination in a straightforward manner." The master had clearly made a favourable impression on the judge, who tended to believe him.

- Competence of master and crew

  As explained above, the master had sailed with the Company for a number of years prior to the incident. Consequently, he was familiar with the way the Company ran their ships. As explained, he had sailed as chief officer on board *Torepo* and therefore knew the ship very well before he took command. The cadet who was acting as lookout had three years experience at sea. One week before the incident a vetting survey had taken place on behalf of an oil major. In addition to other things, the inspector noted: 'competence of personnel is no problem.'

- Vessel's equipment

  The only item of equipment to which the claimants drew attention and alleged not to be working was the echo sounder. No evidence was produced to substantiate that allegation. On the other hand the ship operator was able to produce documents to show that the equipment had been overhauled in dry dock only a week before the incident occurred. A certificate had been issued by the classification society confirming that there were no outstanding items and a detailed report of the dry docking had been prepared by a technical manager from the ship operator's office.

- Procedures and systems

  The oil major vetting inspector who had attended the vessel the week before noted in his report: 'the standard of record keeping was very good with everything readily available. Operational procedures on board were good.' The master wrote up standing orders and night orders and generally ensured that the Company policy was being followed. The officers of the watch seemed to be aware of the Company policy and also read and signed the standing orders and night orders. A navigational procedures manual was in use on board along with the Bridge Procedures Guide published by the International Chamber of Shipping. Proper passage plans had been prepared to cover the possible alternative routes that the vessel might take. When the vessel sailed from Buenos Aires she did not have the correct charts on board for the Patagonian Channel. However, these were brought on board by the local pilots who had been engaged to advise the master during that passage. The passage plan and general exchange of information between master, pilots and officers of the watch did take place. Prior to the incident the chief officer had been monitoring progress on passage, including plotting the vessel's position on the chart at regular intervals and using parallel indexing techniques on the radar to check distances off the nearest land.

The Incident

At 0600 hours on the day of the incident, the vessel was proceeding on her passage through the Patagonian Channel. Everything was proceeding smoothly. The master was in his cabin resting. He had left night orders to the effect that he should be called if the officer of the watch was in doubt about anything or otherwise needed the master. He was intending to return to the bridge at about 0800 hours when the ship would be transiting a particularly difficult part of the channel. On the bridge was the chief officer, one of the two local pilots (the second pilot was also resting), a helmsman and the cadet officer/ lookout. During the course of a few short minutes a whole series of mistakes were to happen.

A major alteration of course point was being approached. The pilot went to mark the ship's position on the chart and to transfer the position onto the next chart. As he did so he realised that the latitude and longitude positions did not coincide with the position of adjacent land and islands. In fact there was a difference of one mile. There is nothing particularly unusual about this sort of thing happening — often the surveys on which original charts had been drawn took place many years before and precise positions may have been difficult. However, this was sufficient to distract the pilot's attention for a few minutes. The chief officer realised that the alteration position was being approached but by this time, presumably, had developed a certain amount of confidence in the pilot and had assumed, wrongly, that the pilot must have been delaying the commencement of the turn on account of some local current or similar. The cadet/lookout saw a light open (become visible) but did not report this. In fact this was the light that should have indicated to the pilot when he should turn. By the time the pilot realised that he had overshot the alteration position the vessel was closing quite rapidly on an island dead ahead. If an attempt was made to alter course at that stage then there would be a serious risk of ripping the side out of the ship which could result in loss of life or personal injuries, an explosion/fire, pollution, loss of cargo and possibly loss of the ship. Instead the correct course of action was taken, apparently without panic. As much speed as possible was taken off the vessel and she was driven straight onto the island. The greatest strength in the ship is in her bows where there are stiffening frames plus a collision bulkhead.

The vessel ran aground at a speed of about six knots. No one was injured, there was no explosion or fire, there was no pollution and no loss of cargo. There was damage to the ship structure at the forward end but a salvage tug was able to pull the vessel off the island safely, with no further loss or damage. In his conclusions the Judge stated:

"the claimants have failed to establish that the casualty was occasioned by causative unseaworthiness. Their claim accordingly fails."

The witnesses and the documentary evidence were sufficient to satisfy the judge that, although relatively old, this was a well run ship with a company and crew trying hard to implement and follow good practices i.e. to make their SMS work. The incident occurred ahead of the Phase One ISM compliance deadline and preparations were still being made, therefore, to have a fully verified ISM system in place. The author is in little doubt, having studied the judgement, that the judge did have ISM principles very much in mind when evaluating the evidence and considering what might be the correct level of operational practice.

What had happened in this case was a series of mistakes, human errors, which all occurred at the same time in the same place. They were errors in the navigation/management of the vessel. There was no evidence to indicate that similar errors or mistakes were a regular feature on board this ship. Rather, what happened was a most unfortunate sequence of mistakes. Whilst these are the author's words and not those of the judge, it is clear that whilst human beings are employed on board ships they will, occasionally, make mistakes. To err is human! Neither the courts nor the ISM Code expect perfection since mistakes will be made. We must use these mistakes as learning opportunities to make sure similar things do not happen in the future. Provided everything else indicates that those involved are trying hard to implement proper safe systems then they should not be punished because of an isolated mistake.

Of course it may not always be possible to exonerate someone who has made a mistake, particularly in cases of strict liability, a pollution incident for example. However, even in such cases, if a ship operator is able to produce witnesses and documentary evidence that, ordinarily, they have very good pollution prevention procedures in place and these are very carefully followed — such that the present incident really was a one off — then the fine that may be imposed is likely to be at the lower end of the scale.

Case study 3 — *The Patraikos 2*

This case was heard in the High Court of Singapore. The relevant facts of this particular case are quite short and whilst this cannot accurately be described as an ISM case it does have a number of very important ISM related implications.

Patraikos 2 was a large container ship that ran aground on what is probably the most powerful lighthouse in the southern hemisphere — Horsburgh Lighthouse. The vessel was aground for 103 days and then taken to Singapore for repairs. She remained in Singapore for another three months. Repairs cost approximately US$4 million. Most of the lost time would probably not have been insured.

The cargo owners claimed for various categories of loss and damage to their cargo as well as their contribution to the general average/salvage costs. They claimed that the vessel was unseaworthy at the commencement of the voyage on account of the utter incompetence of the second mate, who was navigating the ship at the time of the incident. The ship operators argued that they were entitled to rely on the error of navigation as a defence under the Hague-Visby Rules.

The cargo claimants went further and argued that the vessel owner had failed to exercise due diligence in checking on the background, training and qualifications of the second mate before he was appointed. The ship operators were unable to produce any meaningful procedures in which their recruitment policy was set out. Nor were they able to produce any evidence to show that they followed good and safe practices in the recruitment of the second mate in question.

The second mate was called as a witness in the High Court and the judge had this to say of him. *"(he was) clearly incompetent, whose testimony in the witness box shifted like the sands when washed by the tides, depending on his moods"* [see owner of the cargo laden or lately laden on the ship or vessel *Patraikos 2* v the owners of the ship or vessel *Patraikos 2*, Admiralty in Rem No 81/96, High Court of Singapore, 9 May 2002].

The court held that the vessel owner failed to exercise due diligence in checking on the background, training and qualifications of the second mate before he was appointed. As such, the vessel owner was not entitled to claim the error of navigation defence.

ISM Lessons to be learnt

These three case studies should help put into context the full significance of ISM implementation. They should clarify a number of the misunderstandings and misconceptions that arose during the research behind this book. By a process of extrapolation from the cases, the author suggests that the following are the major issues to come to light:

- The court will scrutinise the SMS and the documented systems very closely. Anyone who believes they can get away with only paying lip service to ISM is seriously mistaken.
- It is unacceptable for a ship owner/ship manager to dump ready made procedures manuals on board a vessel and just expect those on board to get on with it. If any company or individual believes that this applies to them then they need to consider their position very carefully and implement major corrective action immediately.
- Voluminous and particularly irrelevant manuals are likely to be criticised by the court. If any company or individual, on reflection, believes that their procedures and checklists and such like would be criticised by a court because of this reason, then immediate corrective action needs to be put in place.

- The recruiting, vetting, familiarisation and training of sea staff is likely to come under very close scrutiny, particularly when an incident occurs because of some human error. A company needs to examine its own human resource systems and procedures to satisfy itself that they are tight enough to withstand detailed interrogation should an individual master's or seafarer's actions lead to a court case, whether in a civil or a criminal action. What evidence is likely to be available to prove that proper vetting, familiarisation and training procedures were indeed followed on each occasion?

- The evidence required will not be limited to the particular incident. Rather, evidence will need to be produced that good procedures were in place and that, for most of the time, those procedures were being correctly implemented and followed. In other words, that the incident that had arisen was an unfortunate accident, an isolated incident, a one off. The Company needs to ensure that proper records are kept, in such a form that they could be produced in court if needed.

- Neither the courts nor the ISM Code expect perfection — people will still make mistakes. However, provided the ship operator can demonstrate that they had good systems in place and all involved were doing their best to properly implement their SMS then they should not be punished or penalised unreasonably by the courts. The extent to which a company or an individual can prove this will depend almost entirely on good accurate records having been maintained.

- Although not specifically addressed in the court cases under consideration, it is suggested that the court will not only expect certain things to be produced as evidence but will also expect audit trails to be capable of being followed. The audit trails should prove whether or not the correct procedures were being followed. The audit trail should bring to light what went wrong with the system and, consequently, why the incident occurred. Reports of similar incidents – whether they be accidents, hazardous occurrences or non-conformities will be examined – as well as the corrective action taken will be scrutinised. If such reports exist and the proper corrective action was taken, even though another incident has still occurred, these reports should help the ship operator's defence. If the claimants/plaintiffs produce evidence of previous incidents which had not been reported then this will be seriously prejudicial to the ship operator's defence — since it would demonstrate that there must be acute problems with the SMS itself.

The view of the author, based on a study of these recent court cases, is that the court recognises that accidents will still occur. If a ship operator has made a proper commitment to ISM, including motivating personnel implementing the systems who can demonstrate they are doing their best, then ISM will be the best friend they could possibly have. If the ship operator has bought an off the shelf system or otherwise put on board an unmanageable system and/or failed to motivate personnel into implementing the system, then ISM will be the worst enemy they could ever imagine.

In an attempt to bring this matter into some sort of context, consider this next report from an Indian chief officer who described, in a very honest and open way, the meaningless rituals he goes through to suggest on paper that the SMS is being implemented, whereas in practice it is just a fudge or facade. He describes the situation this way:

It is no doubt that ISM has improved working standards in the ships, but still the paperwork that ISM requires (especially our Company wants a work-plan, work done weekly, bimonthly and monthly) most of the time we tend to fill up the papers rather than actually checking things or verifying. In this process, I feel there is no use of ISM if people are going to find loopholes in this system and still do things in their own way. I believe this paper work req by the Company is little too much that many times I make work plans for those work which never existed and again make work-done reports for those work done (which actually were not) this is because of fast turnarounds. Since I work in Indian company, and also come to know that many of the Indian companies follow the same procedure. The very basic safety policy of the Company if you see will be the same for most of the companies whereas it should be specifically made for each companies by the shore staff. Even the words are so same, many companies just copy the safety manuals from some other company's and sometimes forgetting to erase that company's name and put their name. I sincerely hope that this situation will be changed soon and something will be done about this, otherwise like the IMO chief said "don't make ISM just a paper work" will be continuing.

A number of obvious and logical questions arise:

- Does the ship operator actually know that such practices are taking place?
- Does the ship operator approve of or condone such practices?
- Has the ship operator tried to persuade the seafarers to behave in a different way?

Presumably audits would pick up the irregularities. Clearly such practices go to the very root of the SMS and even though DOCs and SMCs might exist as pieces of paper they are meaningless. Indeed, such a SMS must be considered seriously deficient and non-compliant. The potential consequences for the ship operator could be very serious indeed. If an incident did occur which necessitated professional investigators to look at what may have led to the incident and they uncovered the practices being revealed by the chief officer, it is the ship operator who would have a lot of questions to answer. A court or inquiry would almost certainly conclude that, if causative, the ship was unseaworthy and the ship operators had failed to exercise due diligence to make the vessel seaworthy. The ship operators would probably lose their Hague-Visby defences and maybe even their insurance cover. It would be no excuse to say that they did not know what was going on onboard their own ships, their system should have brought the irregularities to light at an early stage.

There are still many issues which the courts need to address in connection with ISM to demonstrate the full significance and potential consequences to a ship operator and to those involved in the implementation process, in failing to comply with the requirements of the ISM Code. There are two closely connected scenarios which the author anticipates will be issues to be addressed by the courts in the not too distant future:

1. The question of whether a ship owner, found to be in serious breach of ISM requirements, will still be entitled to rely on the right to limit financial liability on a tonnage limitation basis or be exposed to unlimited liability?

2. The question of whether a ship owner has breached the terms of the insurance cover, H&M or P&I, as a result of serious non-compliance with the requirements of the ISM Code. Could a ship owner find that there is exposure to a very large claim and that there is no insurance cover?

In both cases it is quite possible that the role of the designated person and, in particular, the implied state of mind of the highest levels of management, will come under very close scrutiny and will need to be considered by the court.

In addition to the above two civil actions it can easily be imagined that if another major pollution incident occurs, or worse still a major passenger ship or ferry incident with loss of life, the ISM systems will again come under very close scrutiny by the criminal courts and corporate manslaughter is a very real possibility. A whole host of other criminal sanctions, potentially involving very large fines and lengthy prison sentences, await those convicted. The ship owner and senior levels of management, as well as the designated person, are probably exposed as well as the middle managers and superintendents in the office and the master and officers on board ship. This will clearly vary from one jurisdiction to another. The point is, though, that the court will look at the way in which the ISM systems had been set up and implemented. For the very same reasons as those identified in the civil action cases, the ISM Code will be the best friend a company or an individual could possibly have or the worst enemy they could ever imagine.

Each company and each individual needs to look carefully at the ISM system with which they are working, even though they may have the DOCs and SMCs proudly displayed on the bulkhead and ask themselves whether it complies with the following mantra:

- Say what you do.
- Do what you say.
- Show that you have done what you said that you do.

If the answer is an unhesitating yes, then fine. Such a company or individual probably has nothing to worry about. If the answer is no, or there is some hesitation, then very serious thought needs to be given to consider what corrective action is needed, without further delay.

Chapter 6

THE NEED FOR
AN INTERNATIONAL SURVEY

Introduction

As suggested earlier, little if any empirical evidence seemed to exist that confirmed, or otherwise, whether the ISM Code was working, was starting to work or was having any tangible effects. Before starting the research project the author had started to glean an impression, from talking to serving masters and officers as well as shore based ship operators, that there were some who were speaking in quite favourable terms about their experiences of ISM implementation and some who seemed to have had bad experiences and were very strongly opposed to the whole idea.

It is worth reflecting on a few of the varied views in order to consider what we are really up against. A typical encouraging comment was submitted by a ferry master:

"The crew are definitely more safety aware and basic training greatly enhanced."

Some seafarers had recognised that not only was it proving useful in itself but they had found that they could use the ISM Code to their advantage. As one master put it:

"ISM has made things better. If you use the system properly it puts the pressure back on the office, as they have to be seen to be putting things to rights."

Others would then start to paint a quite different picture. Some seafarers seemed to have had quite contrary experiences — as one master reported:

"The only thing that ISM has changed is the volume of paperwork. Now instead of doing planned maintenance we only have time to write about what we should be doing. Instead of training we now have checklists. The blame mentality is still there now there are thousands of pieces of paper to make sure the finger gets pointed to some poor soul and not the accountant who starved them of the funds to do things properly."

One second engineer did not require many words to sum up his experience of ISM when he stated quite simply that his view of the code was:

"Waste of paper and time!"

For others there were various degrees of resentment; they felt that they had had the ISM Code inflicted upon them. Indeed there were many masters and officers, with whom the author discussed ISM, who stated most forcefully that they had had good safety systems in place for many years, did not need this formalised system and certainly resented the additional paperwork that seemed to accompany ISM as unwelcome baggage.

What was becoming apparent from these discussions was that different people were reporting very different experiences. Sometimes a likely explanation was forthcoming when it transpired that they were working for very different types of ship operating companies. However, there were also examples of individuals working within the same company who had very different views, perceptions and experiences of whether the ISM Code was working or not. It was perceived, therefore, that if an understanding of the current status of implementation was to be established then the views of a very wide range of individuals would have to be obtained.

To establish the current status of implementation

It had been recognised at an early stage that even though the survey questionnaires could be completed on-line, the fact was that many seafarers on board ship would not have access to the Internet. It was therefore decided to print paper copies of the questionnaire for distribution to seafarers.

Initially it was considered that a survey of members of The Nautical Institute would provide a good sample. With 38 branches around the world and over 7,000 members in 70 countries, that should provide a good overall picture. However, it was then recognised that restricting the survey to members of The Nautical Institute could produce quite a distorted picture. Firstly, by definition, the sample would be almost exclusively masters and deck officers. Secondly, such a survey would exclude a very wide range of seafarers who, for various reasons, had not joined the professional body or otherwise did not have access to the Institute journal *SEAWAYS*. Whilst The Nautical Institute membership was most definitely to be included, it was decided to expand the scope of the survey to include as wide a range of seafarers as possible.

The Institute of Marine Engineers, now the Institute of Marine Engineering, Science and Technology (IMarEST) carried a feature article in their journal *MER*, directing their members towards the dedicated website where they could participate and complete the questionnaire on-line. The Merchant Navy Trade Union NUMAST not only carried a major article in their *Telegraph* - but also agreed to enclose a copy of the questionnaire – a circulation of about 25,000! Similarly, the International Federation of Shipmasters' Associations also distributed questionnaires with its own newsletter.

However, whilst the individuals likely to be contacted through these sources would now include engineers and other officers, the survey was still limiting the potential individuals being contacted and consulted to those reading the newspapers and journals of the professional bodies based in the U.K. Other sources of distribution of the questionnaires had to be found to include those seafarers who might not have the opportunity of reading such publications.

Whether or not as a result of divine intervention or inspiration will have to remain a point for speculation but the author approached the Mission to Seafarers to see if they could help and received a most positive response. Two allies and supporters were found within the headquarters of the mission in London — the Reverend Canon Ken Peters and the editor of the mission newspaper *The Sea*, Gillian Ennis. In addition to a major feature article which appeared in *The Sea* and the inclusion of a further 25,000 copies of the questionnaire, Canon Peters wrote to all the chaplains in the missions around the world asking them to take copies of the questionnaire with them when they visited ships in their ports. The chaplains were asked to encourage as many ranks and nationalities of seafarers as possible to participate and complete the forms, assisting them with the task if necessary. Many of the mission stations are, in this 21st century, linked up to the Internet and therefore the chaplains were also asked to encourage seafarers to utilise that facility, to visit the dedicated website and to complete the forms on line.

Batches of questionnaires were also sent to over 300 nautical training establishments around the world, with a request that the head of faculty distribute the questionnaires to mariners who might be ashore studying for their professional qualifications.

Most of the shipping newspapers and magazines carried feature articles about the research project and included a request for their readers to visit the dedicated website and complete the appropriate questionnaire on-line. *Lloyds List* even created a direct link from it's own site to the ISM research site. Completing the questionnaires on line would have proved a considerable help to the author since the data would then have been dropped automatically into the relational database which had been set up and which would be used to analyse the data in due course. Data from the paper questionnaires required manual input. However, it became apparent that relatively few questionnaires were being completed on-line compared with the dozens of completed paper questionnaires that quickly started to arrive from masters and seafarers. It was therefore decided to print and distribute paper copies of the versions of the questionnaire for the ship operators and other stakeholders.

Whilst some ship operator versions of the questionnaire were sent directly to specific shipping companies, batches were sent to the national ship owners associations with a request that they distribute them; making their own request of their members to complete and return the forms and participate in the survey. Some organisations were extremely helpful and supportive. Ship owners organisations were also approached and again some were very responsive and helpful. BIMCO gave considerable coverage to the project in their own publications and created a web link from the front page of their website to the dedicated ISM site.

Clearly, the two key players in ISM implementation were going to be ship operators on the one hand and seafarers on the other. However, there are many other individuals and organisations directly or indirectly involved in ISM implementation who could provide valuable objective assessment of the current status of implementation — as viewed from their particular perspective. A non-exhaustive list of the types of individuals and organisations anticipated to fall into this category includes:

Agents, charterers, classification societies acting in their capacity as a classification society, classification societies acting on behalf of a Flag State Administration, Flag State Administrations, H&M underwriters, insurance brokers, ISM consultants, lawyers, maritime college lecturers, P&I insurers, P&I representatives, Port State control inspectors, pilots, press, professional bodies, ship brokers, ship owners associations, surveyors/consultants, trade unions and university lecturers/academics.

A UK pilot explained the role that such an independent third party could play in providing an impartial objective evaluation of apparent implementation:

"Working as a pilot (mainly tankers) I am not directly involved in the operation of the ISM Code. Boarding ships of various owners and nationalities gives a good opportunity to observe standards on board and get the views of masters

and officers. Since the ISM Code was introduced there has been no noticeable improvement in standards. We see the same ships and the same people on them, all that has changed is that ships staff are further burdened by a mass of paperwork. The success of the code seems to depend on companies operating within the spirit of the code but the companies which really needed the code are hardly likely to enter into this spirit and just see the code as another bureaucratic obstacle to overcome or circumvent."

By obtaining data and views from this potentially very wide range of individuals and organisations, both seagoing and shore based, directly and indirectly involved in the implementation process, a fairly clear picture should emerge of the current status of ISM Code implementation.

Identifying what is going right and what is going wrong

From informal discussions and the various articles that had appeared concerning the status of ISM implementation, it was anticipated that some of the results were going to be fairly predictable. There seemed to be widespread criticism of the amount and irrelevance of much of the paperwork and checklists which masters and senior officers were now expected to complete, supposedly as part of ISM. There also seemed to be a considerable reluctance to report anything other than the most serious accidents which one couldn't avoid reporting, because of various fears and apprehensions.

Fortunately many seafarers are practical and helpful individuals who want to comply and do a good job in a professional manner. Sometimes the starting point is to recognise that things might not be going quite as planned and that steps do need to be taken to rectify the situation, as one Indian able seaman pointed out:

"If ISM Code has to be only concerned with paperwork i.e. making reports then it has been achieved. But if it is concerned with safety then there is a lot of things to be done in this regard."

The survey questionnaires were constructed to address these and many other issues. The three versions of the questionnaires were not identical but were integral and, on the whole, addressed the same issues. The intention was eventually to bring all the data together to compare and contrast opinions between different sectors and to construct the big picture — the belief being that only by viewing the big picture could a real appreciation be made of what was going right and what was going wrong.

Inevitably, the structure of the questionnaires received criticism from different parties, particularly those who perhaps approached ISM implementation from one specific direction. An example of such criticism can be seen in the following observation received from a Port State control inspector:

"Your questionnaire is rather restricted to accidents and non-conformity reporting. In my opinion it is more useful to check, during PSC inspections, how the ship's crew is working, how the managing Company is backing up the vessels; is the Company really interested in upgrading maintenance and indeed actively following the ships maintenance/monitoring the ships maintenance. Some parts of the ISM Code for the companies/vessels are solely obligatory parts to receive the DOC/SMC, so they make only the minimal obligatory effort: many times a SMS is far substandard."

The issues addressed by the PSC inspector are very important. It is hoped that many of these issues were addressed by the respondents in the narrative section of the questionnaires if not in their answers to the questions contained within the questionnaire. Unfortunately the questionnaires already included about 30 questions on six sides of paper and, if the document increased any further in size, there was a very good chance that few people would have the time or would take the trouble to complete it. The author extends his apologies to those who would have liked other issues covered or the emphasis changed, but trusts that many of those issues will actually come out through the numerous comments and observations from respondents which appear in this book.

A number of respondents provided detailed comments. Some of these are felt by the author to contain such important, first hand, experiences that they are reproduced in full or with a minimum amount of editing. One such report, most relevant to this section, where we are making an initial appraisal of what is going right and what is perhaps going wrong, was submitted by a British master who explained that he had set up safety management systems on board ships in five different companies and consequently had considerable relevant experience to share:

"There is still widespread fear among crews and officers that this is all directed against the vessels and crew. The concept that the owners/manager becomes partly liable has been lost. There is acceptance that extra paperwork is inevitable, but no faith that it will make any difference. The time lost to administration of SMS is thus not often balanced by a gain in method and control. From the level of SMS compliance I find on assuming command of vessels, many masters are at best not understanding and at worst ignoring ISM, and audit is obviously not

finding it. In other companies (not here) I am often asked to go onboard vessels and tidy up SMS just before audit (I refuse). ISM is capable of being a good tool, but at the moment it is too large. It has become a means to legitimise an unscrupulous operator and burden the well-intentioned."

Whilst the experiences of those directly involved in the implementation process are perhaps the most relevant of all, the observations of independent third parties can be extremely valuable. This is enhanced considerably when coming from an individual who has seen ISM from a number of different perspectives such as the following Port State control inspector:

"As an ex-seafarer who has seen the ISM Code from all sides; as marine superintendent, ship's master and as a regulator — I firmly believe that though ISM in itself is a good thing and if used properly will improve the safety standards of vessels — it will never work as long as the guidelines are so generic and all encompassing. Also many companies are using ISM and its procedures as an excuse for non investment in comprehensive training. To make the system work the guidelines should be more descriptive. I carried out a PSC on a bulk carrier classed with a major society and it's ISM manual was generic for the whole fleet which included tankers, bulkers and RoRos. I think ISM should be for individual classes of vessels not generic for fleet."

There are a number of very important issues raised by the PSCI which will, perhaps, not receive unanimous support but do deserve a response. In this section of the book, however, the intention is merely to flag up various issues and explain the approach which has been taken when conducting the survey and research. Detailed responses to these and other issues raised in these introductory sections appear later in the book.

It often takes an external observer to adopt a dispassionate view and raise some fundamental questions, such as the one raised by a maritime university lecturer:

"If the SMS is working, why do we have all other controls, e.g. oil companies vetting inspections, etc? They don't seem to trust the system."

There were many seafarers amongst the respondents who saw ISM as attacking their professionalism or it was generally perceived to be a cause for the decline of all that had been good at sea. A small selection of some of these are reproduced below providing a flavour of those perceptions whether we personally agree with them or not:

"ISM has created more work. ISM undermines the professionalism of the engineers engaged in their duties. ISM allows paper engineers to shine and

professionals to be considered out of date. ISM is all about paperwork. Paper maintenance is easy to achieve." (British chief engineer)

No-one seems to trust people – at one time we had Lloyds or whoever and Flag States. That was enough. Then we got vetting, ISO 9000, ISO 9001 now ISO 14000 + Audits + ISMA + ISM + Port State control. Who is going to be next to check that the rest are doing their jobs? Someone has lost the plot. We are supposed to get ships from A to B — maintain them, get the cargo in and out safely etc. Now we fill in forms. It has accomplished nothing but increased a few jobs and a lot of bullshit." (British master)

"ISM is too bureaucratic. Paperwork rules the job rather than being its servant, and has become a database upon which lawyers can build their same old arguments. I believe ISM has contributed to the decline in morale at sea. It has suppressed individuality, with the ship owner, in practice, being no more accountable than he ever was." (British ferry master)

"Crew standards + level of training continue to decline." (British chief mate)

"Increased workload on master and chief engineer involving ISM implementation negates half of what is meant to be achieved i.e. significant man-hours lost every day just filling in forms, filing records and checklists etc. I presently sail with an all British crew and feel that the system is wasted. Previously I sailed with all foreign crew and the system, once understood, had some significant effect." (British master)

"Filling up checklists has become an integral part of today's job. However, I do feel that all officers in general and junior officers in particular need to practice good values of seamanship. This would reduce lots of accidents/near misses etc." (Indian master)

"ISM ignored human factor — too tired of paper work!!" (Korean chief officer)

"There is too much stress on documentation taking most of the time. Attention is more towards correct maintenance of documentation than some times actual practice being carried out. If documentation is reduced slightly actual implementation of ISM will be more significant." (Indian chief mate)

Again there are numerous important and controversial issues being raised in these examples. The view of the author is that many of them are based on very serious misunderstandings of the very basic principles of ISM, and are cause for serious concern. Many of the issues are addressed and explored fully in later sections.

For many masters and seafarers and indeed for ship operating companies, ISM was their first encounter with formal management systems. In some cases it appears that little if any training or preparation was provided to explain the principles behind the idea of such management systems. Consequently, misunderstandings at that most basic level were introduced at a very early stage and it would appear that, in some cases, these have still not been addressed. A classification society auditor based in the Middle East explains the problems, as he has experienced them, when attending on board vessels:

"The ISM Code has created a paper chase on board vessels which fails to address the main problem — crew training. Some companies assume that by having a auditable system that complies is a substitute system for good quality crew. Audits are carried out in accordance to their documented system — ISM — just as similarly a QA audits a company — it doesn't imply it's good i.e. because a chocolate factory complies to its QA system it doesn't mean the chocolate tastes good. Similarly, a ISM system on a vessel doesn't mean the vessel is any safer or good – it just complies to a auditable system – does it?"

Another class society auditor reported similar experiences and also highlighted the important role of another new concept, the idea of a safety culture:

"ISM is a quality system. To work in a quality system it is necessary to: understand what a quality system is — believe in the quality system; be educated to the safety culture. Safety culture is something which cannot be learned from books."

Perhaps not surprisingly, since the fundamental concepts of formal management systems might not be fully understood or appreciated, the idea of a cycle of continual improvement that should be at the heart of such systems is also lacking. Part of this understanding involves identifying hazardous occurrences, near misses, non-conformities and the like; reporting them, analysing them, finding out what has gone wrong with the system, learning lessons and implementing corrective actions. In this way the cycle of continual improvement helps to make people involved and the ship and the Company safer and more efficient.

There are of course many other players in related industries who are watching very carefully how ISM is progressing. The insurers — both hull and machinery and P&I — clearly have a vested interest. Surprisingly, few cargo insurers seem to make statements openly about ISM. Whilst they are perhaps a little removed from the immediate operation of the ship, they clearly have a direct interest in the quality and standards of ships being used to carry the cargoes they are insuring as well as the people working on board those ships and the people ashore operating those ships. One H&M insurer, based in the Mediterranean, had clearly given the matter some very careful consideration and shared his thoughts on the matter:

"The concept of ISM is obviously a first class idea. To the very good companies it made no difference as essentially all ISM was doing was formalising what was their normal operational practice. To other companies ISM was perceived as a

bureaucratic burden and was operated with reluctance. ISM will only operate well if the owners wish it to. On some vessels the volumes provided by the owners fill whole shelves and are clearly never read or used. A conscientious attempt should be made to limit the number and size of the ISM manuals. The lack of continuity of service by crews who are generally supplied by agencies is not conducive to the efficient operation of ISM. ISM has to be implemented on board as a way of life, but in reality on most vessels it is perceived as yet another paperwork burden that further limits the time available to actually operate the vessel. If there is a delay to the vessel or cargo operation due an ISM defect then it is expected that this would be held against one or more members of the crew and could lead to disciplinary action or dismissal, hence the reluctance to report defects." (H&M Insurer — based Mediterranean)

Suggesting what needs to be done to move forward in a positive way to improve safety management on board ships

Having progressed from establishing the current status of implementation to identifying what is going right and what is going wrong, the logical next step is to consider whether any practical suggestions could be made to those companies, ships and individuals who might be experiencing difficulties to move forward in a positive way to improving the management of safety on board.

As the completed questionnaires began to arrive along with their narrative comments it started to become clear that there were some who had gone through such an experience and had either emerged from, or at least were coming out of, the other side with a new found conviction that ISM could work and that it could be a most useful tool for helping to manage safety on board. An Indian master shared his experience as follows:

"ISM system is now transforming from paperwork culture to implementation culture. On board systems have been organised a lot since its implementation. Most operations are now carried out in a predetermined and planned way. It has given a systematic approach and laid down minimum safety standards to be followed. ISM system has streamlined all ship related operations on shore and ship."

An Australian master reported a very similar experience:

"Initially there was a lot of confusion associated with the ISM system but as time has progressed so has the system and it is now reasonably user-friendly. Parts are under constant revision to ensure the safest and best practices exist at all times."

Some respondents who were positive about the progress being made had no hesitation in making a clear link between the attitude of the

84

Company running the ships and successful functioning of the SMS such as the following Indian chief officer:

"It is how your company implements the code which is crucial in deciding how honestly and effectively the objectives of the code are achieved. In my own experience I feel that shipboard and shore based personnel are now viewing ISM in a more positive and serious manner and not an unnecessary pain any more."

These types of reports are clearly most encouraging and many other seafarers could probably relate similar experiences. A European Flag State Administration had no doubts when it reported:

"The ISM Code is working and will improve the situation over time."

Other seafarers appear to have had most unfortunate experiences and it may take a considerable amount of persuasion to convince them that even if these masters and others are correct then ISM can work. The paperwork can be brought under control and made relevant and the ISM Code can make a significant and positive contribution not only towards the way safety is managed on board but the way the whole ship operation is managed. In just a few words this British chief officer perhaps encapsulates the frustrations experienced by many:

"The only difference the ISM Code + ISO 9002 has made is a huge increase in paperwork and hence working hours."

Many others appear to be at an earlier stage of transition from being sceptics and the comments of this Australian second engineer are quite typical of many which were received:

"I believe it will take more time, a culture cannot change over night. We the seafarers are trying and I am sure those results will show. A simpler documentation system is needed as the paperwork has become a nightmare. These SMS are massive document driven software programmes and we need time to get to know them."

The respondents not only flagged up the problems but many also suggested ways in which progress could be made with implementation. Some of the suggestions were based on their own learning experience and have the potential for being most useful in guiding others who are at the early stages of the learning process to help them avoid some of the pitfalls. Sometimes the advice is very simple and basic but address issues which are perhaps all too often ignored, or forgotten about, or just taken for granted. A classic example was provided by a Filipino master:

"Communication between ship to shore is most vital importance in achieving ISM Code a success."

Indeed, as will become evident in later sections of this book, one of the major factors which appears to be inhibiting the successful implementation of the ISM Code in some companies is an inadequate level of communication between ships and the office ashore. In some cases this goes far beyond just a failure to keep in touch with each other, to out-and-out mistrust, perceptions that there is no support or interest, to feelings of isolation and abandonment. In some cases it was possible for the author to see the perceptions from both sides — to see the comments coming from seafarers and those coming from management ashore. It was of considerable concern to see that each side had very different perceptions to the other. The message each had been trying to convey to the other had been lost somewhere along the way.

Often it was impossible not to share in the absolute despair expressed by some masters, who had tried very hard with ISM implementation but ended up feeling as though they were trying to push water uphill. A British tanker master expressed such despair:

"Attempts made on the vessel to comply with ISM Code have met with little or no response from Company and when response is received is mainly dismissive. The other master — when I am on leave — does not bother with ISM and Company does not follow this up."

An issue which became apparent from numerous responses from seafarers was a perception that all shore management are using ISM for is to cover their own backs. A significant number stated quite clearly that in their view the shore management were using the ISM Code to shift all the responsibility and liability from the office onto the ship. If such perceptions were in fact true it would of course represent a very serious situation. What is equally worrying, though, is that such perceptions demonstrate a fundamental misunderstanding of the ISM Code. Anyone who has read the Code could not be left in any doubt at all that the obligation and responsibility for the implementation and operation of the Code rests with the Company and that responsibility is non-delegable. This issue is looked at closely in due course, as well as other significant misunderstandings and myths that seem to have developed. Typical of the misinformed type statements is the following from a British master:

"The ISM Code has shown senior management to be doing all that is possible to run a safe operation. Should prevent them being prosecuted. Is the operation really safer? I do not believe so."

The Company has the responsibility of not only developing the structure of the SMS but also of bringing it alive within the Company. If a ship operator did believe that they could get away with only implementing the former then they would be seriously misleading themselves. Indeed they are likely to have a very serious shock coming their way if problems arise in bringing the SMS alive that result in injuries or damage. Some third parties also seemed to share a somewhat jaundiced view of certain categories of ship operators. A P&I correspondent put it this way:

"I am located in a part of the world where profits rather than safety are the single biggest consideration in the operation of ships generally. Ship operating companies see the ISM Code generally as just another piece of paper which is required to keep ships running. They do not participate in the true spirit of the code, preferring instead to do the minimum to obtain certification. Sorry I can't be more positive!"

Another fundamental issue which started to become apparent was the crucial importance which should be attached to the employer/employee relationship and in particular the need to have a loyal and familiar crew. In other words, issues relating to continuity of employment as highlighted by a Norwegian fleet manager:

"The key issue in respect of the ISM Code is stability of crew"

The failure to have such stability or continuity amongst the crew almost certainly results in the crew having no meaningful sense of ownership of the SMS. Without such ownership it is difficult to understand where the motivation would come from to make the system work. Without such motivation it may have to be conceded that the most magnificently written procedures manuals in the entire industry are little more than a few thousand words typed on pieces of paper. The significance of this is perhaps reflected in the following observation from a British master:

"ISM is applicable to company systems. People operate these systems. Due to continual change out of personnel retraining in ISM is only done on board informally as and when items surface. The code is only as good as the personnel. If no training and no continuity is achieved the objectives will never be met."

For many seafarers the biggest problem area raised was the enormous increase in paperwork they had encountered, which they attributed to the ISM Code. Whether implied or stated explicitly, clearly this would be a most important issue to address if significant progress was to be made with implementation. An Indian chief engineer linked the necessity to the recognition that there also seemed to be a constant stream of ever more rules and regulations coming forward with which ships and operators had to comply:

> *"With impossible number of various regulations by various regulatory bodies has made the owners/operators totally sceptical about the practicality of these ISM Codes and has made them comply with these regulations not by choice but by fear of losing (the trade or business). Whereas it was the intention of the ISM Code to have the culture of safety consciousness from within and did not need to be forced upon. So time has come to sincerely review its effectiveness by all the concerned parties involved with shipping, one thing although is sure that the present situation of excessive paperwork/checklists/forms must be done away with if any long term good is expected from this ISM Code."*

There seemed to be a recognition by some respondents that certain ship operators, at least, did require policing and also recognised the important function Flag State and Port State authorities may have to play in that policing role. As a Bangladeshi second engineer remarked:

> *"ISM implementation body should create continuous pressure on shipowners, managers, manning agents and marine industry related personnel for implementing the codes to maintain safer sea life."*

It also became apparent that many respondents seemed to be in little doubt that standards differed considerably with ISM implementation, not only between ship operating companies but also between Flag State Administrations, classification societies and Port State Administrations. Numerous first hand examples were provided which seem to confirm that these are very serious issues which need addressing. With such perceptions in mind it is understandable that some seafarers and ship operators are despondent, to say the least. One British chief engineer clearly felt very strongly about the significance of such inequalities:

> *"I find it worrying that very different standards exist between different issuers of DOCs and whilst some ship owners have expended a great deal of time and money to produce a viable ISM system, others have produced something that is totally inadequate. Until this dual standard is rectified, honest ship owners are at a considerable disadvantage. I do not know why I and my officers should spend a good deal of valuable time on a minimally manned ship completing paperwork that has little if any value, does not make this vessel any safer or more efficient than prior ISM. I would urge that ISM is consigned to the dustbin of history where it deserves to be — unlamented and unloved."*

Some respondents seemed to have very clear ideas of where the blame lay for the apparent failure of ISM. Whilst they had suggestions about what should be done they were perhaps less than constructive in their suggestions for finding solutions. The opinion put forward by one British master provides a good example:

"Best to get the present laws enforced before bringing in new ones. ISM was a good tool but has been corrupted by incompetent third world substandard class and flag authorities. Time to get the IMO act together and sort this unholy mess out."

An interesting general overview of what, in the opinion of this particular contributor, was going wrong and some suggestions on what might need to be done to put things right, was put forward by a surveyor working in the Far East:

"Broadly speaking I consider the ISM affected companies as follows:

- *Good/first class operators who never needed ISM in the first place have implemented fully and is operating well.*

- *Medium class operators. Typically they asked experts to set up the system. The experts set up an over-complicated system that is almost impossible to implement effectively. Little or no benefit has been gained except that all ships have a list of designated persons to contact in case of emergency.*

- *The cowboys. To them it is just another certificate. Typically, ship's staff do not take an active part in ISM. All reports are filled in after prior consultation with head office (the head office tell them which section has to be completed). The shipboard manuals are never read. The shipboard manuals are generally too big and complicated. In some cases the crew cannot even understand the language in which they are written. Ship's staff complain about the amount of paperwork involved.*

- *Too much time spent in form filling and too little on the job. It can at times be dangerous for ship's staff to have to give priority to form filling over and above the safe and efficient execution of their duties. How many big accidents will be caused by form filling will remain to be seen, but some day people will realise the negative impact of safety on form filling and hopefully do something about it.*

- *Checklists are futile. A checklist is only as good as the man and pencil ticking little boxes. If the man is competent why does he need a checklist? It would appear that the ISM Code is assisting operators employ cheap low skilled crews with the benefit that they can get adequate on board training with checklists. Senior personnel have a difficult job onboard running ships with these inexperienced check list crews who invariably have not received proper shore based training.*

- *The implementation of the code on substandard ships, typically eastern bloc, is leading to the shifting of much tonnage to recognised managers. I expect the reputation of some of the leading management companies will be damaged by this shift. Finally I have three recommendations for the future of ISM:*

  - *Simplify it down to the bare safety necessities.*

  - *Make operators responsible for ensuring that the crew they employ fully understand and can operate the on-board system before they join the vessel.*

Contained within the last opinion are many controversial, indeed radical, ideas and proposals. This is the case with many of the comments that have been received from masters, seafarers, ship operators and many others involved in the industry who responded to the survey and provided their valuable input. It is many of those observations and proposals that are examined in more depth in later sections of this book. However, at this juncture the background and underlying concepts behind the ISM Code are considered in more detail. It is suggested that a good grasp of these issues is crucial before the analysis of the survey and a full appreciation of the extensive comments and suggestions received from the respondents can be made.

Chapter 7

PARTICIPANTS IN THE SURVEY

Introduction

This chapter examines who was invited to participate in the survey and who actually participated. It describes a profile of the respondents — showing that they represent a very wide cross section of the international shipping and related industries and professions.

The questionnaires were designed to include three categories of participant:

1   Masters and other seafarers (blue)

2   Ship operators (green)

3   Other stakeholders (red)

Each group is considered in turn and later chapters compare and contrast the responses from the different category groups. Specific details of individuals, ships or companies were not asked for and, although in some cases these details were given, it was decided to maintain total anonymity throughout the survey. The general view was that personal details such as names were not a significant factor in the investigation. However, it is conceded that it would have been useful to compare and contrast the views of sea staff and shore staff in the same Company. This was possible in a few instances. In general terms, as will be seen, there do appear to be quite significant differences of perception regarding ISM implementation between those working ashore compared with those on board ship. It also became apparent that there are significant differences of perception between different individual experiences.

Masters and seafarers

Completed questionnaires from masters and other seafarers started arriving very soon after the initial distribution with *SEAWAYS* and the NUMAST *Telegraph* in April 2001. Perhaps not unexpectedly, the majority of those early responses were from British masters and senior officers, either shore based, or involved in the short sea trades or offshore supply boats. Alarm bells started to ring when those responses were reviewed, since many were suggesting quite a negative attitude towards ISM with few words of support. Indeed, as the second wave of responses started to arrive from Australia, New Zealand, Canada and the USA a similar general pattern seemed to be developing – although a few individuals were starting to surface who were showing a much more positive attitude. Eventually more and more completed questionnaires were being returned — from Indian, Filipino, Eastern European masters and officers — and a much better balance of opinion was starting to take shape. As expected, most seafarers' responses were received as completed paper copies, although a significant number completed the questionnaire on line and left some most valuable and interesting comments and observations on the discussion page of the website.

A number of ship owners were very supportive of the project and agreed to encourage the active participation of their seafarers. Typically, a supply of questionnaires was sent to the master of each ship in the fleet with a request that the master encourage everyone on board to complete one. On a number of occasions the master was specifically asked to encourage everyone to be frank and honest and to send their completed questionnaire direct to the author. The intention, of course, was to reduce to a minimum the risk that individual seafarers might be concerned there would be repercussions if they gave the wrong answers and if those answers were seen by the management ashore. Other companies made no such suggestions and the questionnaires were returned via the ship operator's office — although it was emphasised to the master that each officer and crew member should complete the questionnaire independently. A review of all questionnaires coming from the same ship held additional interest. On a number of occasions the perceptions of the seafarers were remarkably close to each other. Whether this reflected some level of cooperation between everyone on board in completing the questionnaire or otherwise genuinely reflected the way in which ISM was working on board that particular ship was difficult to judge. The participation of those ships and in particular the support from the Company ashore and the master on board, was very much appreciated.

Further bundles of questionnaires started to be returned from nautical colleges and other training academies around the world, providing a most useful input from seafarers of nations who were outside the initial distribution group. It was not always easy to determine, but on occasions it was very clear that completed questionnaires were being received from seafarers through the Mission to seafarers in various parts of the world.

Nearly 2,000 completed questionnaires were received from the seafarers of at least 54 different nations. Not unexpectedly, the majority of responses were from the masters and senior officers, although a significant number were received from junior officers and ratings. Many of the responses from junior officers and ratings contained extremely valuable comments and useful observations.

Position on board — masters and seafarers

The options of categories of seafarer provided were based on a very traditional style manning arrangement and proved adequate for the vast majority of, but not all, respondents. For statistical purposes the categories were grouped together as follows:

• Masters.

• Senior Officers — Chief Engineer, Chief Mate, Second Engineer.

• Junior Officers — Second Mate, Third Engineer. Third Mate, Junior Engineer, Other Officer.

• Ratings — Petty Officer, Senior Rating, Junior Rating.

Masters 36% 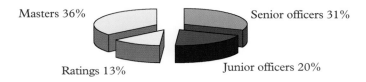 Senior officers 31%

Ratings 13%   Junior officers 20%

Figure 7.1 — Seafarers who responded to the ISM survey

Initially the responses were mainly from masters — though a significant number of completed questionnaires were appearing from chief engineers. Even by July 2001, when a preliminary analysis of the figures was undertaken, almost exactly 50% of the seafarers respondents were masters and the majority of them were from OECD countries. However, as time went on, more and more of the completed questionnaires being returned were from other ranks and other nationalities.

The number of masters — representing 36% of seafarer respondents and senior officers at 31% may be disproportionate to their actual numbers in relation to other seafarers on board. However, it is perfectly understandable that these ranks in command and responsible for the on-board implementation of the SMS were the ones most prompted to complete the questionnaire.

What was very encouraging was the significant contribution from more junior officers and ratings. It is certainly correct to state that the SMS directly involves everyone on board. Rarely, though, has the author seen or heard any views expressed by these categories of seafarers on ISM implementation prior to this survey. The ability to identify the different ranks on board allowed a comparison to be made of perceptions of the ISM Code and the working SMS between those categories of seafarers. It was also possible to analyse and compare their responses by national groups.

Nationality of masters and seafarers

The first draft of the questionnaire included a request for respondents to declare their nationality for two main reasons — firstly, to help complete the profile of the individual seafarer and secondly, to try to ensure that a good and fair representative sample of seafarers had participated in the survey. The real significance of including this information was not to become fully apparent until well into the survey when one of the biggest surprises of the exercise was to manifest itself. This is discussed in a later chapter. Responses were received from nearly 2,000 seafarers from many different nationalities, as shown in the following table:

Algerian	German	Pakistani
Australian	Ghanaian	Panamanian
Bahamas	Greek	Polish
Bangladeshi	Icelandic	Portuguese
Belgian	Indian	Romanian
Brazilian	Irish	Russian
British	Italian	South African
Bulgarian	Jamaican	Spanish
Canadian	Korean	Swedish
Chinese	Kuribatan	Swiss
Croatian	Latvian	Syrian
Danish	Lithuanian	Taiwan
Dutch	Malaysian	Tanzanian
Ethiopian	Maldivian	Thai
Filipino	Maltese	Turkish
Finish	Myanmar	Ukrainian
French	New Zealand	USA
Georgian	Norwegian	Yugoslavian

Figure 7.2 — Table showing the nationality of respondents to the survey

94

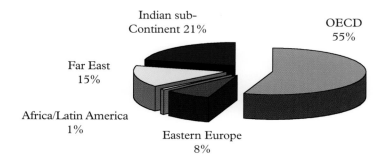

Figure 7.3 — Nationality of seafarers — ISM survey sample

Length of service with ship operator — masters and seafarers

Monumental changes took place in the shipping industry during the late 1970s and 1980s. Ships from traditional maritime nations were flagged out in considerable numbers and many seafarers from those nations were replaced by personnel from countries where labour prices were lower. Prior to that it was quite common, particularly for officers, not only to enjoy long term company contracts of employment but also remain very loyal to their company, often spending their entire career at sea with the same Company. That loyalty and consequent bond between employee and employer was very strong. It tended to engender considerable professional pride that, in turn, contributed to a positive attitude towards the safe and efficient running of the ship. With the breaking of those bonds there were inevitable consequences, evidenced by the enormous rise in accidents and claims that occurred during the middle to late 1980s.

It was, therefore, considered appropriate to try and establish through the questionnaire whether there might be any correlation between the length of service of the individual seafarer with a particular company and their attitude towards the ISM Code and the working of the SMS. Interestingly, a significant number of the seafarers who responded were long serving staff with the same Company.

The seafarer respondents were almost equally split between what can perhaps be considered short to medium term employment — with the same Company for up to five years and what can be considered long term — above five years. It was a pleasant surprise to the author to find that 29% of respondents had been in the employment of the same Company for more than 10 years.

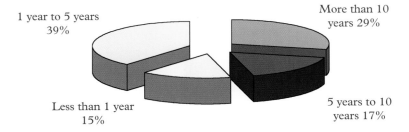

1 year to 5 years
39%

More than 10
years 29%

Less than 1 year
15%

5 years to 10
years 17%

Figure 7.4 — Length of service with present company

A relevant comment was received from the chief executive of a Scandinavian shipping company who said:

"Crew from manning agents not always taking company goals and objectives to their heart because they will be gone after a few trips. Lack of ownership because short employment. Others with long term employment recognise however the importance of a functional SMS because it makes their job easier by providing routines and the basis to train newcomers"

It became apparent that a key factor always present in those companies where ISM had been successfully implemented was continuity of employment — particularly the seagoing staff.

Type of ship — masters and seafarers

It was considered important to establish, through the survey, what type of vessels individual seafarers were serving aboard. Phase One of the ISM Code implementation, with the final deadline of 1st July 1998 for compliance, involved all passenger ships including passenger high speed craft, oil tankers, chemical tankers, gas carriers, bulk carriers and cargo high speed craft of 500 gross tonnage and upwards. The deadline for Phase Two implementation was set for 1st July 2002 when all other cargo ships and mobile offshore drilling units of 500 gross tonnage and upwards had to comply.

The survey took place during 2001, providing an opportunity of looking at the experiences of Phase One ships and hearing views from those preparing for Phase Two implementation. In addition to establishing the type of vessel on which seafarers were serving, the survey also established whether the vessel held a Safety Management Certificate (SMC). It became apparent that many Phase Two ships falling into the category of other cargo ships — particularly container ships, refrigerated cargo ships and offshore supply boats — had gone

96

through the verification and certification process well ahead of the 1st July 2002 compliance deadline.

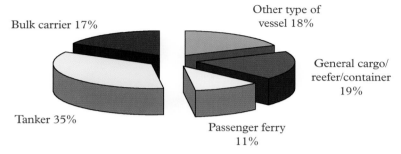

Figure 7.5 — Types of ship — seafarers

Initially there was a disproportionate amount of responses from ships that fell into the category of Other Types of Vessel. These comprised, primarily, naval support type craft, survey vessels and offshore supply type vessels. This was probably because those seafarers were the first to receive their questionnaires. Eventually, a much more balanced profile of the typical ships at sea was achieved, providing the potential for a good representative sample of views and opinions to be obtained.

Size of ship — masters and seafarers

In order to extend and elaborate on the general profile of the vessels on which the sample of seafarers were serving, the questionnaire also allowed size and age of ship to be considered.

These size and age results complement the results showing the types of ship — suggesting that the sample of seafarer respondents were serving onboard many different types and sizes of vessel — thus providing the propensity for a good cross section of views being considered.

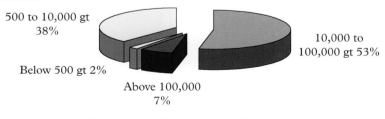

Figure 7.6 — Size of ship — seafarers

Age of ship — masters and seafarers

To complete the profile of the ships on which the sample of seafarer participants were serving, the questionnaire asked about the age of the vessels.

10 to 15 years 18% 15 to 20 years 23%

Below 10 years 35% Above 20 years 24%

Figure 7.7 — Age of ship — seafarers

The profile of ships according to age was fairly well balanced either side of the 15 year mark. Because the SMS is primarily concerned with the human element and management side of the ship operation, an analysis of the significance of age of ship on frequency of accidents/claims was not undertaken in this study.

Size of fleet — masters and seafarers

It was felt important to try and ensure that responses were being received from staff of shipping companies of different sizes. Whilst there are still a few large fleets in private hands, most large fleets probably fall into three broad categories:

- Oil majors/liner operators.
- Large third party ship management companies.
- National state fleets.

These categories of ship operator were probably much more likely to have developed and implemented formalised safety management systems well ahead of any ISM requirements.

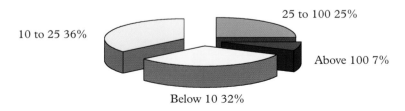

25 to 100 25%

10 to 25 36% Above 100 7%

Below 10 32%

Figure 7.8 — Size of fleet — seafarers

Interestingly, there were only about a third of seafarer respondents who probably fell into this category. The other two thirds seemed to be sailing with small to medium sized companies. This indicates that the cross section of the sample was from a good variety of types of company background.

Corporate structure — masters and seafarers

The conclusions reached based on the size of the Company were reasonably well supported when the corporate structure itself was examined.

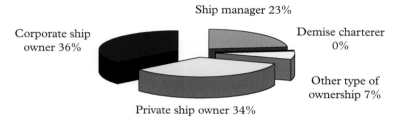

Ship manager 23%

Corporate ship owner 36%

Demise charterer 0%

Other type of ownership 7%

Private ship owner 34%

Figure 7.9 — Corporate structure — seafarers

Seafarers in the other type of ownership category would appear to be working primarily for national government agencies — either in naval support type vessels or national fleets.

Management system background — masters and seafarers

During the 1970s and increasingly so into the 1980s, most of the oil majors started to implement quality management type systems. A number of the large ship management companies were not far behind – particularly those who were members of the International Ship Managers Association (ISMA). These were all voluntary, or at least not mandatory, schemes and were often verified against International Standards Organisation (ISO) quality assurance (QA) standards.

As part of their own QA standards it was often a requirement that any 'supplier' or 'subcontractor' would also need to be QA accredited. Accordingly, as time went on, more and more ship owners realised that if they were going to continue to tender for charters from these operators then they would have to go down the QA road. The author is surprised, though, at the very high proportion of respondents who claimed to have had a formalised QA system ahead of ISM.

99

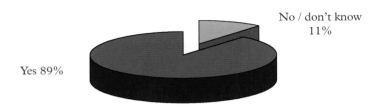

No / don't know
11%

Yes 89%

Figure 7.10 — Existing QA system — seafarers

The most probable explanation is that many ship operating companies developed a QA system alongside their ISM systems during the run up to Phase One implementation. One of the implications of this, though, is that individuals working within those companies should already have been well familiar with formalised, documented, management systems, as well as such things as non-conformity reporting. The indications from other parts of the survey suggest that this might not be the case.

Flag of vessel — masters and seafarers

Many of the criticisms levelled at the decline in standards within the shipping industry start with the concept of Flags of Convenience (FOCs) or perhaps Open Registers. Seafarer's organisations — in particular the International Transport Workers Federation (ITF) — have long waged a campaign against the very idea of FOCs, alleging that they have allowed the living, working and remuneration standards of seafarers to be eroded. This also extends to safety standards. A detailed inquiry into many related issues was conducted by the International Commission on Shipping, under the Chairmanship of the Hon. Peter Morris.

Much valuable work has been done by organisations such as the Seafarers International Research Centre based at Cardiff University in an attempt to quantify the extent to which such claims may be valid. It is probably fair to say that there most certainly are instances where particular FOCs have lived up to the reputation painted of them by the ITF. However, it should also be stated that some FOCs/ Open Registries conduct their activities to very high professional standards.

Within the context of this survey it was considered appropriate to try and establish whether seafarers were working on board ships flying

100

their own national flag or a FOC and, if so, whether there was any apparent connection between that fact and their attitude towards ISM.

An analysis of the flags showed a variety almost as extensive as the nationality of the seafarers themselves, as the list below shows:

Algerian	German	New Zealand
Antigua and Barbuda	Gibraltar	Norwegian
Australian	Greek	Panamanian
Austrian	Hong Kong	Papua New Guinea
Bahamian	Indian	Philippines
Barbadian	Indonesian	Portuguese
Belgian	Iranian	Qatar
Belize	Irish	Russian
Brazilian	Isle of Man	Saudi Arabia
British	Israeli	Singaporean
Brunei	Italian	South African
Cambodian	Jamaican	St. Vincent and
Canadian	Japanese	Grenadines
Cayman Islands	Jordanian	Swedish
Chinese	Kuwait	Swiss
Cypriot	Liberian	Thailand
Danish	Luxembourg	Turkish
Dutch/Netherlands	Malaysian	Ukrainian
Ethiopian	Maltese	United Arab
Falkland Islands	Marshal Islands	Emirates
Fijian	Moroccan	United States
Finish	Myanmar	Vanuatu
French	Netherlands Antilles	

Figure 7.11 — Analysis of flags

It is not always easy to be accurate in identifying when a flag is a national flag or a FOC. For the purpose of the analysis undertaken here it was decided to accept the list of the ITF which identifies 30 countries as having so-called flags of convenience:

Antigua and Barbuda	Barbados	Bolivia
Aruba	Belize	Burma
Bahamas	Bermuda	Cambodia

101

Canary Islands	Honduras	Netherlands Antilles
Cayman Islands	Lebanon	Panama
Cook Islands	Liberia	St. Vincent
Cyprus	Luxembourg	Sao Tome & Principe
Equatorial Guinea	Malta	Sri Lanka
German International	Marshal Islands	Tuvalu
Ship Register	Mauritius	Vanuatu
Gibraltar		

Figure 7.12 — The ITF list of flags of convenience

Using the ITF list of so-called flags of convenience a comparison can be made of the respondents who were sailing on national flag vessels and FOCs:

Flag of Convenience
48%

National Flag
52%

Figure 7.13 — The split by flag

This is probably not far away from the true profile of the international fleet — although the survey sample indicates a slight bias towards the national flag. This was probably as the result of relatively large participation by British seafarers sailing onboard UK registered ships.

Ship operators

Within this category the survey was looking to identify ship owners and ship managers rather than chartering organisations. It was looking for the Company who had actually set up and was operating the SMS — within the context of ISM that meant the Company with the big 'C'. Attempts to reach this category of potential respondent was threefold:

1   Direct mail — both by post and by e-mail.

2   Through national ship owners associations.

3   Through specially targeted editorial in shipping magazines, newspapers and journals.

There are so many individual ship operating companies around the world that available resources would not allow a wholesale direct mailing approach. However, it was possible to identify about a hundred ship owning and ship management companies with a significant number of ship units in their operation and limit the direct mailing to those companies. There were also a number of companies with whom the author had already established contact and had an existing dialogue.

In many maritime countries with a ship owning industry, the industry has formed trade associations or chambers of shipping to provide a voice for itself and generally promote the industry as a whole. Most shipowners of any significance are members of their national association. Most of these national shipowners' associations are themselves members of the International Chamber of Shipping and/ or the International Shipping Federation. It was therefore possible to rationalise the mailing a little by sending a letter to the secretariat of each individual association or chamber, along with a supply of questionnaires, asking for their help in distributing the questionnaires to their members and encouraging participation.

There are also other, more specialised, ship operator organisations who were also potential sources of help with contacting ship operating companies. The largest is perhaps the Baltic and International Maritime Conference (BIMCO), based in Copenhagen. The main work of BIMCO is in drafting and regulating a whole range of standard shipping contracts such as charter parties, bills of lading and similar. BIMCO had already been proactive in providing training and familiarisation with the ISM Code and had also conducted a limited survey of its own members. BIMCO were extremely helpful with the research, including feature articles in their own newsletter to their members about the project, encouraging them to participate, as well as a direct link from the front page of their own website to the author's own ISM website.

Intercargo is an organisation of dry cargo ship operators. They too offered a lot of help and support and became directly involved at a later stage with the production of the *Seafarers Guide to ISM* that is discussed later.

Most ship operating companies around the world subscribe to one or more of the leading shipping newspapers and/or magazines — specifically *Lloyds List* and *Tradewinds* as far as newspapers are concerned and *Fairplay*, *Lloyds Ship Manager* and *Seatrade* as far as magazines and periodicals are concerned. By providing the editors/ journalists with an interesting and maybe a little provocative or controversial interview or article, the author could almost guarantee prominent editorial coverage which would reach the attention of the ship operators around the world. With the cooperation of the editors and journalists it was also possible to include a personal request to ship operators to participate in the survey and provide them with the relevant contact details, including the website address. *Lloyds List* in particular were most kind and went one step further by displaying a scrolling banner on the front of their own website asking their readers to participate in the survey and provided a link direct to the ISM website.

Through these various sources it was possible to reach a very significant proportion of the ship operators of the world.

Position in the Company — ship operators

Within any ship operating company almost everyone has some level of involvement with the ISM system — though some will clearly be much more involved than others. The level of involvement for individual positions within shipping companies may vary considerably between companies. This may be because of the way a particular company has established its systems but, probably, is more directly dictated by the size of the Company. For example, in a very large oil major or liner operator there may be a number of individuals who devote their entire time to overseeing the SMS in their capacity as Designated Person (DP). In a more modestly sized company the operations manager may also double up as the safety manager and also deal with the insurance and claims matters. They may also be expected to fulfil the role of the DP.

It was hoped that a good cross section of views could be obtained from different groups of people with a variety of ship operating organisations and this seems to have been achieved.

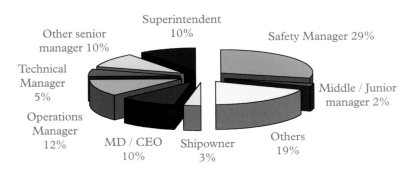

Figure 7.14 — Respondents — ship operators

104

It is made very clear in the preamble to the ISM Code, paragraph 6, that 'the cornerstone of good safety management is commitment from the top'. It was perhaps a little disappointing, therefore, that there were not more responses directly from the ship owners, MDs and CEOs. Having said that, the author was in direct contact with a number of individuals at that very senior level and it became very apparent that the cornerstone prophesy of the architects of the Code was fully borne out.

It was also of relevance, perhaps, that the individuals most interested in participating in a survey such as this would be the DPs. The ISM Code does not itself set out who exactly within the Company should undertake the role as DP, only that such a person should have direct access to the highest levels of management. In an ideal situation the DP should be independent of a line management function but should have sufficient seniority to ensure that he/she can properly fulfil the role of DP. This would include ensuring that the SMS is functioning properly and that all safety aspects are being adequately resourced and supported.

Analysis of the responses indicates that two thirds of those submitting responses on behalf of the ship operator categories were DPs.

No 35%

Yes 65%

Figure 7.15 — Designated Person

What is not clear though is whether, or to what extent, the DP doubles up in some other capacity within the ship operating Company.

Nationality of respondent

Not every respondent declared their nationality – although less than 10% declined. The nationalities represented in the sample included:

Australian	British	Cypriot
Bangladeshi	Bulgarian	Danish
Belgian	Canadian	Dutch
Brazilian	Croatian	Filipino

Finnish	Jordanian	Romanian
French	Korean	Russian
German	Mauritius	Singaporean
Greek	Mexican	South African
Hong Kong	Moroccan	Spanish
Icelandic	New Zealander	Swedish
Indian	Norwegian	Swiss
Indonesian	Pakistani	Turkish
Iranian	Peruvian	Ukrainian
Israeli	Polish	United States
Italian	Portuguese	

Figure 7.16 — Nationalities of respondents

It is important to recognise that the question asked for the nationality of the individual. Whilst many of the individuals were working inside their own native countries there were many more who had taken their expertise elsewhere and were part of an expat-type labour force. In this regard, compare the above list with that displayed in figure 7.27 below.

Length of service with company — ship operator

Compared with sea staff, shore based respondents tended to have been in the employment of the same ship operator for a much longer period of time. Indeed, 43% claimed to have been employed by the same Company for more than 10 years and another 25% for more than five years.

5 to 10 years 25%

More than 10 years 43%

1 to 5 years 24%

Less than 1 year 8%

Figure 7.17 — Length of service — ship operators

What is not clear from the responses is how much of that time was actually spent in employment ashore and how much might have been spent working in the same Company but at sea. It has been a tradition for very many years for ex-seafarers to be employed within the industry ashore and, where possible, many companies have preferred to recruit from within their own staff.

106

To some extent the issue is not necessarily relevant, since the purpose of the question was to establish to what extent continuity of employment might be a factor in attitudes towards ISM and in the successful implementation of a SMS.

Types of ship — ship operator

In addition to asking which type of ship was operated, the questionnaire also asked the respondent to indicate the number of each type of ship operated by that Company. The graph below is based on the percentage of the total number of ships identified.

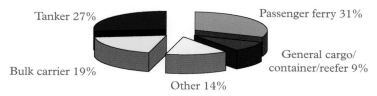

Figure 7.18 — Types of ship — ship operators

We see a good cross section of ship types. Not surprisingly the majority are Phase One ships — passenger, tankers and bulk carriers. There was also a not insignificant number of Phase Two ships — though many of those appear to have achieved verification and certification early.

Sizes of ship — ship operator

Ship sizes were broken down into four broad categories — small ships, below 500gt; medium size ships between 500 and 10,000gt, large ships between 10,000 and 100,000gt and very large ships above 100,000gt. Whilst a small number of respondents were operating ships below 500gt, most were operating ships in the medium to large range, with a strong leaning towards ships between 10,000 and 100,000gt. There were no respondents operating very large ships. The respondents appear, therefore, to operate the most usual size of deep-sea vessel.

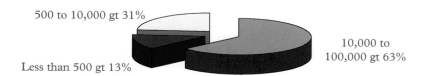

Figure 7.19 — Sizes of ship — ship operators

Age of ships — ship operators

The age profile of the vessels represented by the ship operators in the survey did not vary enormously from that of the seafarer group – with a fairly even balance each side of the 15 year mark. Again this sample would seem to be quite representative of the world fleet as far as age is concerned.

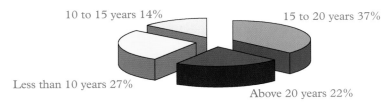

Figure 7.20 — Age of ships — ship operators

Size of fleet — ship operators

A quarter of the respondents appeared to work for relatively large ship owning or ship management companies but the vast majority were engaged with small to medium sized operators — operating up to 25 units.

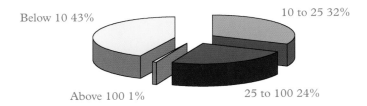

Figure 7.21 — Size of fleet — ship operators

Corporate structure — ship operators

From analysing the actual questionnaires it would appear that some respondents had difficulty ticking the correct box here since they seemed to operate both as ship owners in their own right as well as ship managers. What appeared to be the most appropriate category was chosen. However, there did appear to be a somewhat disproportionate number of private ship owners although bearing in mind that most of the fleets were in the small to medium size bracket — this is probably correct.

108

Private ship owner 40%

Corporate ship owner 25%

Demise charterer 1%

Ship manager 34%

Figure 7.22 — Corporate structure — ship operators

Management system background — ship operators

There was quite a significant difference between answers given by seafarers to those provided by shore side staff of ship operators when declaring the extent to which the Company had previously been involved in formal QA type systems. One explanation might lie in the understanding of what constitutes a formal QA type system. Another possible explanation is that the ships were subjected to QA type systems — particularly in the lead up to Phase One implementation, compared with the office ashore who might already have been familiar with such systems. In any event, the survey still suggests that the majority of respondents from ship operator offices did have previous knowledge/experience of working in management type systems.

No 29%

Yes 71%

Figure 7.23 — Prior experience of management type systems — ship operators

Flags of vessels — ship operators

There were surprisingly few different flags represented by the respondents:

Antigua	Greece	NIS
Bahamas	Hong Kong	Panama
Barbados	India	St. Vincent
Brazil	Italy	Sweden
British	Korea	Turkey
Canadian	Liberia	UAE
Cyprus	Malta	United States
Dutch	Marshall Islands	

Figure 7.24 — Flags represented by ship operator respondents

109

Again the sample was split almost equally between companies operating ships flying their own national flag and those using FOCs/Open Registries. The sample would therefore appear to represent a reasonably accurate cross section of the international industry as far as registry is concerned.

Flag of Convenience 43%

National Flag 57%

Figure 7.25 — Flag — ship operators

Main centre of operation

The other area of interest, as far as ship operators were concerned, was where their centre of operation was based. This produced a quite different list of countries:

Bahamas	Greece	Portugal
Belgium	Hong Kong	Singapore
Brazil	India	Spain
Britain	Italy	Sweden
Canada	Jordan	Switzerland
Cyprus	Korea	Turkey
Denmark	Mexico	UAE
Finland	Monaco	United States
France	Netherlands	
Germany	Norway	

Figure 7.26 — Ship operators main centre of operation

In terms of the number of companies from each country and represented within regions – the result is as per the following graph:

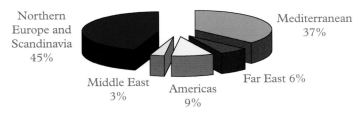

Northern Europe and Scandinavia 45%

Mediterranean 37%

Middle East 3%

Americas 9%

Far East 6%

Figure 7.27 — Ship operators by region

This shows that there does seem to be a bias towards northern European/Scandinavian ship operators, with a significant under representation of ship operators from the Far East.

Other stakeholders

Because of the diversity of individuals and organisations falling within this category it was difficult and particularly labour intensive trying to contact them. Whilst some might have been picked up through The Nautical Institute and other distributions and possibly through other media coverage — it was considered necessary to make personal, direct contact and to supply the correct questionnaire form. The other stakeholders included a very wide range of individuals and organisations. Some of the more significant are set out below:

Flag State Administrations

These are the national government departments or agencies with responsibility for ensuring that ships flying their national flag, i.e. ships registered in their country, comply with all the relevant rules and regulations and are issued with the correct certification – including ISM certification. Many of these Flag States do not actually have sufficient infrastructure or resources to undertake their obligations and responsibilities themselves. They therefore delegate to a Recognised Organisation (R/O) — usually the Classification Society. Their contribution, therefore, was very important in explaining how they had undertaken the verification and certification process and to describe the types of problems they had encountered.

Each of the Flag State Administrations has a delegate/representative at IMO, although some countries are much more active in their participation than others. Initially, individual/personal letters and questionnaires were sent to each of the 158 member state delegates, care of the IMO address in London. That did not solicit much response, so the exercise was repeated and another full set of individual/personal letters with questionnaires were sent to their mailing address in their home countries. This generated a more significant response, but still many of the major Flag States did not respond.

Port State Control (PSC) Administrations

The PSC authorities are also national government agencies/ departments and attempts were also made to contact them in the same way as those attempting to solicit responses from the agencies handling the Flag State Administration duties. In many cases these departments were one and the same, or at least very closely related. Again, nearly 160 individual letters with questionnaires were sent. The secretariat offices of each of the seven MOUs around the world, including the United States Coast Guard (USCG), were also contacted.

Classification societies

The classification societies were very important potential contacts, since they had at least three possible areas of involvement with ISM implementation:

1   In their role as an actual classification society, where they would be attending vessels in connection with classification matters, thus providing them with an opportunity to observe how the SMS was interacting with maintenance and other class issues.

2   In their role as recognised organisations acting on behalf of Flag State Administrations

3   In their capacity as consultants to companies, where they provided a service setting up the particular SMS.

There are 10 full member societies and two associate members of the International Association of Classification Societies (IACS) and letters and questionnaires were sent to the secretariat of each. In addition, over 600 individual letters were sent to separate branch offices of different societies around the world. Suggestions had been made that there might be some irregular practices taking place in certain classification societies regarding verification and certification and it was considered important to obtain direct individual feedback as well as the party line which might come out of head office.

ISM consultants

Whilst the classification societies almost achieved a monopoly with regard to ISM consultancy as well as verification and certification as ROs — there were a number of independent ISM consultants who did manage to break into consultancy and RO activities. Unfortunately, only about 20 such individuals and organisations were identified. Appropriate letters and questionnaires were sent accordingly.

P&I correspondents

Whenever there is an incident onboard ship that is likely to result in a third party liability claim, the P&I Club will probably be involved. In all major ports and most secondary ports, the P&I Club has a local correspondent, sometimes referred to as a representative. The correspondent attends to assist masters on the spot in dealing with the immediate problem and ensuring that the position of the ship owner and P&I Club are fully protected.

The P&I correspondents, therefore, tend to be at the sharp end of any incident that occurs on board. As a consequence, they have experience of seeing many ships and seafarers in situations where the SMS is under close examination. They are, therefore, in an ideal situation to feed back their experiences of ISM implementation. Through the author's own contact network, individual letters and questionnaires were sent out to nearly 500 correspondents around the world.

Surveyors and consultants

Similarly, whenever there is a H&M or P&I type incident on board ship and indeed in many other situations, surveyors or specialist consultants will be instructed to investigate the incident to establish causation and to evaluate the damage. As such, these individuals, who tend to be very experienced professionals, are in an ideal position to observe how/if safety management systems are working or, if not, what the problems might be. Letters and questionnaires were sent to about 350 individuals and surveying firms around the world.

Lawyers

Following an incident, particularly a serious incident, it is quite likely that a lawyer will be instructed to take evidence/statements, to investigate the matter to establish causation and to prepare the case for fighting in the courts or in arbitration or to enter into settlement negotiations. In a similar way to the surveyors, lawyers are provided with an excellent opportunity to observe how the SMS has been implemented and how it is, or isn't, working. A very handy *International Directory of Shipping Lawyers* is published in conjunction with *P&I International* (Informa) and they kindly provided the author with an electronic version that was most useful for sending a large mailshot of letters with questionnaires to over 500 lawyers around the world.

Insurers

Whilst the P&I correspondents, surveyors and lawyers may be involved at the sharp end of the investigation, their reports are likely to be presented to individuals within insurance organisations. These claims handlers, loss adjusters, managers, underwriters or similar are also being provided with an opportunity of observing the SMS in action, or maybe inaction! A P&I claims handler may have many hundreds of claim files with which he/she is dealing. All the P&I Clubs were contacted with a request to circulate copies of the questionnaire around their claims handlers. Attempts were also made to send letters and questionnaires to H&M and cargo insurers.

Nautical college lecturers

Almost all seafarers spend some part of their career attending a nautical school, college, academy or similar institution. It occurred to the author, therefore, that the lecturers, invariably ex-mariners themselves, would hear from students passing through what they thought about ISM and how the implementation process was going on board their ships. They would also be in a position to make their own assessment as to whether there were any cultural shifts taking place in the attitude of younger seafarers towards safety. Accordingly, letters were sent to well over 300 training establishments around the world.

Pilots

In the vast majority of cases, when a large ship approaches or leaves port they utilise the services of a local pilot, who can advise the master on navigational issues in that port or harbour. In practice the pilot usually takes the ship from the pilot station to its berth. It can be appreciated, therefore, that any one pilot will have an enormous and varied experience of all different types, sizes and nationalities of ship. More importantly they would see first hand how masters, officers and crew — as well as the machinery — work and how the SMS was operating in practice. In addition to a small number of individual letters, a request was submitted to the International Pilotage Association (IPA) asking for help to encourage their pilot members to participate and share their experiences.

Professional bodies and trade unions

Whilst nautical and marine engineering professional bodies, as well as seafarers trade unions and similar bodies, were contacted in an attempt to get the questionnaires to the seafarers, it was also recognised that it would be useful and interesting to have feedback from the administrative and managerial staff of those organisations in order to establish their views and observations.

Others

There were many other individuals and groups who were also contacted, with a variety of backgrounds — as wide ranging as ships agents to marine biologists and conservationists, chaplains and accident investigators.

Who responded? — other stakeholders

In an attempt to keep the illustrations relatively uncluttered, the various individual categories of respondents have been grouped together as follows:

• Service providers — e.g. agents, lawyers, surveyors, consultants, ship brokers, etc.

• Classification societies — in all their various guises.

• Flag state Administrations.

• ISM consultants.

• Port state control — including the various government agencies — as well as the secretariats of the MOUs.

• Educationalists — including college lecturers, academics and other training providers.

• Insurers.

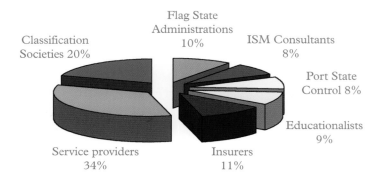

Figure 7.28 — Categories of other stakeholders

The numbers of respondents who submitted completed questionnaires was not great — about 460. However, this group submitted numerous detailed narrative reports, providing considerable insight into the status of ISM implementation as seen through the eyes of third party observers (although clearly some of these observers were very close to the implementation process, e.g. Flag State Administrations and ROs). As a matter of interest and for completeness sake, respondents were also asked to indicate whether or not they were ex-seafarers themselves. They responded as follows:

No 21%

Yes 79%

Figure 7.29 — Other stakeholders — ex-seafarers?

Perhaps not unexpectedly the majority of the other stakeholder category who responded did have a seagoing background.

Geographical base of respondents – other stakeholders

Also for completeness sake it was considered appropriate to establish where the other stakeholders were based, which would perhaps help identify any possible parochial bias which might exist. In fact they came from far and wide:

Australia	Hong Kong	Poland
Belgium	India	Portugal
Brazil	Ireland	Russia
Britain	Israel	Singapore
Canada	Italy	South Africa
Chile	Jamaica	Spain
Colombia	Japan	Sweden
Croatia	Korea	Taiwan
Denmark	Malaysia	Thailand
Finland	Netherlands	Turkey
France	New Zealand	UAE
Germany	Norway	United States
Greece	Philippines	

Figure 7.30 — Geographical base of respondents — other stakeholders

Survey participants – data and information

The completed questionnaires received from the three main categories of respondents have provided an enormous amount of data for the database. This is of considerable importance, but the full potential of this data cannot be realised within the scope of this review. There is an almost limitless number of permutations of queries which can be run on a relational database. Within the scope of the survey a focus had to be maintained on key areas and a restriction had to be placed on the issues considered. It is vital that others who could make good use of it are given the necessary access. There is now also a unique catalogue of detailed comments, observations, reflections and experiences received from nearly 800 individuals, providing an invaluable insight into ISM implementation which again cannot be fully utilised within this review. However, others who have the resources to explore related issues must make use of that data for the benefit of the shipping industry and in particular those who go to sea.

Chapter 8

THE SURVEY RESPONSE ON ISM IMPLEMENTATION

Introduction

What do we mean when we talk about compliance with ISM? This may appear, at first sight, a very simple and straightforward question — it actually turns out to be extremely complicated. We could talk in terms of a process of verification leading to certification. Once the ship operator is issued with a Document of Compliance (DOC) and the ship with a Safety Management Certificate (SMC) we could say — that company and that ship are ISM compliant. The problem is that there is no one, universally accepted, set of standards against which all systems are measured.

The author's view is that the procedures manuals and even the DOC and SMC represent no more than 20% of what is needed to comply with ISM and, at best, provide only a suggestion that the Company and ship have successfully implemented a Safety Management System (SMS). Only once the written procedures and the SMS have become a dynamic, living part of the way things are done will ISM implementation have been fully achieved. What will be necessary to reach such an achievement should not be underestimated. It will take a lot of time, a lot of hard work and a lot of personal commitment by all involved. The end result however promises rich rewards.

SMCs and DOCs

The government of the state whose flag the ship is entitled to fly — the Administration — is responsible for verifying compliance with the requirements of the ISM Code and for issuing the appropriate certificates. The requirements are set out in the new 'Part B — Certification and Verification' of the Code which has been significantly expanded in its scope following amendments in December 2000 by resolution MSC.104(73) – which entered into force on 1 July 2002. Section 13 deals with 'Certification and Periodical Verification':

PART B – CERTIFICATION AND VERIFICATION

13 CERTIFICATION AND PERIODICAL VERIFICATION

13.1 The ship should be operated by a Company which is issued a Document of Compliance or with an Interim Document of Compliance in accordance with paragraph 14.1, relevant to that ship.

13.2 The Document of Compliance should be issued by the Administration, by an organisation recognised by the Administration or, at the request of the Administration, by another Contracting Government to the Convention to any Company complying with the requirements of this Code for a period specified by the Administration which should not exceed five years. Such a document should be accepted as evidence that the Company is capable of complying with the requirements of this Code.

13.3 The Document of Compliance is only valid for the ship types explicitly indicated in the document. Such indication should be based on the types of ships on which the initial verification was based. Other ship types should only be added after verification of the Company's capability to comply with the requirements of this Code applicable to such ship types. In this context, ship types are those referred to in regulation IX/1 of the Convention.

13.4 The validity of a Document of Compliance should be subject to annual verification by the Administration or by an organisation recognised by the Administration or, at the request of the Administration, by another Contracting Government within three months before or after the anniversary date.

13.5 The Document of Compliance should be withdrawn by the Administration or, at its request, by the Contracting Government which issued the Document when the annual verification required in paragraph 13.4 is not requested or if there is evidence of major non-conformities with this Code.

13.5.1 All associated Safety Management Certificates and / or Interim Safety Management Certificates should also be withdrawn if the Document of Compliance is withdrawn.

13.6 A copy of the Document of Compliance should be placed on board in order that the master of the ship, if so requested, may produce it for verification by the Administration or by an organisation recognised by the Administration or for the purposes of the control referred to in regulation IX/6.2 of the Convention. The copy of the Document is not required to be authenticated or certified.

13.7 The Safety Management Certificate should be issued to a ship for a period which should not exceed five years by the Administration or an organisation recognised by the Administration or, at the request of the Administration, by another Contracting Government. The Safety Management Certificate should be issued after verifying that the Company and its shipboard management operate in accordance with the approved safety management system. Such a Certificate should be accepted as evidence that the ship is complying with the requirements of this Code.

13.8 The validity of the Safety Management Certificate should be subject to at least one intermediate verification by the Administration or an organisation recognised by the Administration or, at the request of the Administration, by another Contracting Government. If only one intermediate verification is to be carried out and the period of validity of the Safety Management Certificate is five years, it should take place between the second and third anniversary dates of the Safety Management Certificate.

13.9 In addition to the requirements of paragraph 13.5.1, the Safety Management Certificate should be withdrawn by the Administration or, at the request of the Administration, by the Contracting Government which has issued it when the intermediate verification required in paragraph 13.8 is not requested or if there is evidence of major non-conformity with this Code.

13.10 Notwithstanding the requirements of paragraphs 13.2 and 13.7, when the renewal verification is completed within three months before the expiry date of the existing Document of Compliance or Safety Management Certificate, the new Document of Compliance or the new Safety Management Certificate should be valid from the date of completion of the renewal verification for a period not exceeding five years from the date of expiry of the existing Document of Compliance or Safety Management Certificate.

13.11 When the renewal verification is completed more than three months before the expiry date of the existing Document of Compliance or Safety Management Certificate, the new Document of Compliance or the new Safety Management Certificate should be valid from the date of completion of the renewal verification for a period not exceeding five years from the date of completion of the renewal verification.

The new Sections 14, 15 and 16 deal with 'Interim Certification', 'Verification' and 'Forms of Certificates' respectively. The IMO originally issued a set of Guidelines on the Implementation of the ISM Code by Administrations by Resolution A.788(19). These Guidelines were replaced with Revised Guidelines, which were adopted by Resolution A.913(22) in November 2001. This resolution revokes A.788(19) as of 1 July 2002. Some of the more significant issues have now been taken out of the Guidelines and incorporated into the expanded Section 13 plus Sections 14, 15 and 16 of the amended Code. There are, therefore, three bodies who might issue the DOC and SMC:

- The Administration itself — i.e. the Flag State.
- An organisation recognised by the Administration — often referred to as a Recognised Organisation (RO).
- Another contracting government.

Few Administrations would appear to have sufficient resources, expertise or possibly even the will to undertake their own verification and certification. The United Kingdom is one of the few who are undertaking these functions themselves, through the offices of the Maritime and Coastguard Agency (MCA). As part of his initial research the author wrote to 150 Flag States, care of their delegate who might attend IMO meetings, asking for information about their approach towards verification and certification. Very few responses were received. Further attempts were made to obtain the information but still the vast majority of the Flag State Administrations did not provide any information.

The majority of Administrations appear to have delegated the tasks of verification and certification to recognised organisations. Whilst there are a small number of independent ROs, almost all of the delegation has been to the classification societies.

As far as delegating to another contracting government is concerned, it would appear that the MCA, on behalf of the British government, have occasionally undertaken such work at the request of other governments and have, on even rarer occasions asked other governments to undertake such work on board UK registered ships.

The IMO have endeavoured to encourage Flag State Administrations to face up to their responsibilities but with limited success. For example, MSC.Circ.889 / MEPC. Circ.353 deals with 'Self Assessment of Flag State Performance' along with MSC.Circ.954

/ MEPC.Circ.373 'Self Assessment of Flag State Performance: Criteria and Performance Indicators. It provides Administrations with guidance on measuring their performance and asks them to advise IMO how well they are progressing. As far as can be established, the Administrations who are cause for concern do not tend to participate in such initiatives.

Is the certification in place?

This may appear to be a strange question to ask, but it should be recalled that the survey was originally undertaken between the Phase One and Phase Two deadline dates. The purpose of the question was to provide the possibility of establishing whether there were any different perceptions between those who had already obtained their DOCs and SMCs and those who were still working towards full verification.

In fact almost 90% of seafarer respondents indicated that they already possessed their SMC on board. Only 6% admitted that they didn't know whether or not they had such a certificate. Many of the

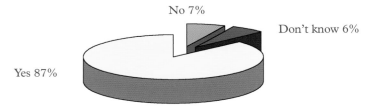

Figure 8.1 — ISM certification — SMC — ships

Phase Two ships on which respondents were serving would appear to have gone for verification and certification early.

A little surprising from the ship operators side was that 16% of the respondents didn't seem to know whether or not the Company had a

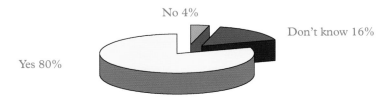

Figure 8.2 — ISM certification DOCs and SMCs — ship operators

122

valid DOC or the ships SMCs. Apart from that, as would be anticipated, most companies did hold a DOC and had SMCs for their vessels.

Who issued the certificates?

Because a significant number of the questionnaires were distributed to British masters and officers through *SEAWAYS* and the *NUMAST Telegraph*, it is probable that a disproportionate number of responses came from people working on board UK registered ships or ships flying affiliated flags. The results received, therefore, may not accurately reflect the true global picture. The responses from the masters and other seafarers indicated that nearly two thirds of all the safety management certificates were issued by the classification societies with nearly one third being issued by the actual Administration.

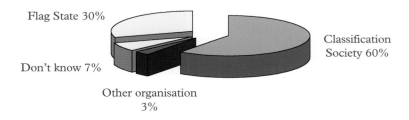

Figure 8.3 — Who issued SMC? — ships

The responses from the ship operators, which also included the issuer of the DOC, mirrored very closely the results from the masters and seafarers with nearly two thirds of the documents being issued by the classification societies and one third by the actual Administrations.

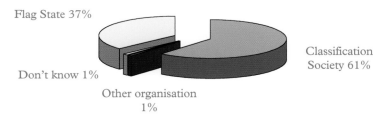

Figure 8.4 — Who issued DOCs and SMCs? — ship operators

Clearly the classification societies hold a very powerful position with regard to verification and certification of very substantial sections of the world fleet. In some respects this may help to introduce a level of standardisation and uniformity into the verification process but on the other hand some caution may be in order when so much power and authority is placed in so few hands.

Other related problems, and causes of possible concern, were raised by various respondents. An engineer shared his experiences:

"I was part of the seven member team tasked with implementing the ISM Code into ferry operation between '96 and '99. We have 40 vessels certificated and have passed our first verification audit from one of the leading classification societies. My concern is that the classification societies are not strict enough when they note non-conformities. This seems a common problem in that the Flag States, for the most part, are not up to speed and do not have their own auditors trained to carry out the verification audit on their own behalf. The classification societies do not want to be too strict with the hand that feeds them, and so we have created a paper tiger, with no apparent claws."

Procedures manuals

Supporting the SMS are procedures manuals and possibly other documents that should be the subject of formal document control. The individual company's SMS should encompass all the requirements of the ISM Code. The structure of the company's documentation should be adapted to suit the Company culture, size and the trading pattern of its ships.

The SMS documentation should consist of both office and shipboard manuals. These manuals should be organised in a manner which allows all employees concerned with the SMS to readily refer to its relevant provisions in the satisfactory performance of their duties.

The Company should ensure that the relationship between the SMS and other shore and shipboard systems are properly understood, and that relevant references and interconnections are established.

It would appear that many companies, for a variety of reasons, have ended up with documented procedures that may be inappropriate and too voluminous.

The SMS documented procedures do not have to be voluminous or overly complicated. The ICS / ISF have provided a possible structure for an uncomplicated SMS documentation system (ICS/ISF Guidelines):

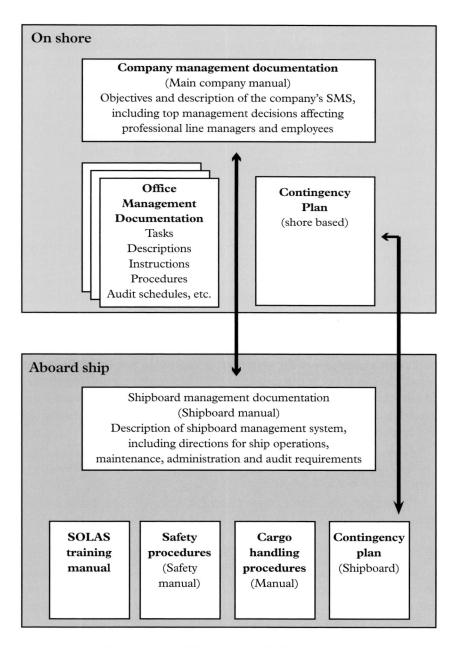

Figure 8.5 — Possible structure for SMS documentation

125

Chapter 5 examined, in some detail, a recent judgement handed down from the High Court in London relating to the vessel *Eurasian Dream*. The ship operator was very severely criticised because of the voluminous, yet inadequate and inappropriate, set of SMS procedures documents they had put on board that ship. This actually contributed to the vessel being found to be unseaworthy and, further, the carrier had failed to exercise due diligence to make the vessel seaworthy.

It is not known just how many of the SMS documentation systems would fall into a similar category to that found on the *Eurasian Dream*. There would appear to be many that are far from adequate. A New Zealand master gives one such example:

"I have recently sailed on a container ship which has had manuals developed entirely from ashore and it was terribly complicated to access information and any non-conformities to do with procedures documented in manuals were ignored due to the fact the author / DPA / safety manager could see no wrong with his work. This created a defeatist attitude with sea staff to not bother to report or seek change."

Clearly, such a system has no value at all and is doomed to failure if the master on board and his officers and crew have no confidence in the documentation. At the end of the day it is those very people who will need to bring the procedures alive through implementation — surely they are in one of the best positions to evaluate the effectiveness and appropriateness of the system. It is suggested that those on board must have their say in the design of the SMS and the implementation of corrective actions within the cycle of continual improvement if the system is to work. A manager from a ship operator's office explained the importance his company put on this aspect:

"I believe the key for us in implementation was to involve sea staff as much as possible in preparing procedures and so on even although this slowed down the process it was worth the effort."

Developing the sense of involvement that leads to the sense of ownership has got to be worth taking the time to achieve properly.

Who prepared/produced the procedure manuals?

Following on from what has just been said, it is of crucial importance to consider who actually produced the procedures manuals. If care has not been taken in the production of the manuals then it is almost certain that the SMS itself will never be successfully implemented. A Port State control inspector has taken the issue to its natural conclusion in two very short sentences:

"Very few safety management manuals have been completed with assistance of ships staff. Most crews feel ISM an unnecessary burden and merely a paperwork exercise."

Stories of off-the-shelf sets of procedures manuals have abounded since the lead-up to Phase One implementation in July 1998. So called ISM consultants were, apparently, offering these ready made safety management systems for sale and merely changed the name of the vessel on the front page of the document. Unfortunately many of the responses to the survey seem to confirm that such practices have been widespread. An Indian master summarised his experiences in the following terms:

"1   A major amount of benefit of the ISM system was to the consultants who made the manuals/drew up the ISM for the system, which in many cases was drawn up without any consideration to the type of operator (i.e. his commitment) — manuals are blindly copied from standard companies. Checklists are filled in just before inspection etc., without actually checking.

2   Though the advantages are obvious (i.e. only if implemented properly and with commitment) unfortunately it is looked upon (still) as a nuisance (due to the time restraints)

3   Commercial pressure compels (there are many more people willing to take your job) you to do things well beyond the scope of the ISM. Though on the first line in the manuals is 'Safety will not be compromised due to commercial reasons' — that is the first thing compromised (read as fatigue / rest hours etc.)

The questionnaire therefore asked for information about who prepared or produced and wrote the ISM manuals on board their ships. The available choices were structured to try and provide an opportunity to select from most permutations.

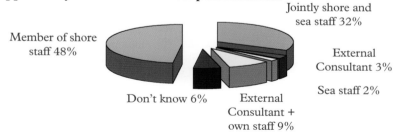

Figure 8.6 — Who produced ISM manuals? — seafarers

It appears that well over a half of the masters and seafarers who responded believed that there was little or no involvement of the sea staff in the production/writing of the ISM procedures manuals and, presumably, in the development of the SMS which had been produced for their vessels. That is clearly serious cause for concern if true. Looking at the positive side though, one third of respondents claimed that it was indeed a team effort between sea and shore staff to develop their ISM manuals and their SMS. It is in that level of participation that the concept of ownership of the system can start to take a hold. Once there is a sense of ownership of the system by those on board and in the office, then a safety culture can take hold and, it is suggested, the full benefits can almost be guaranteed!

The ship operators were asked to consider the same question. Interestingly the results were quite similar with perhaps a little greater emphasis on the thought that the bigger input in the development of the manuals came from shore staff and external consultants rather than the involvement of the sea staff.

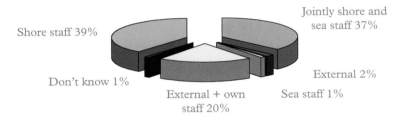

Figure 8.7 — Who produced ISM manuals? — ship operators

A comparison of views on the results of the answers to this question illustrate the similarities and differences of perceptions between the two groups:

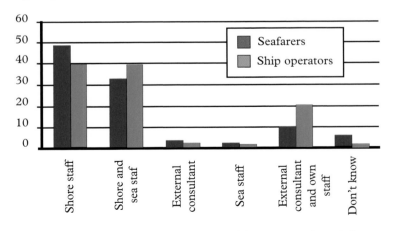

Figure 8.8 — Who produced ISM manuals? — a comparison of views

Does it really matter who produced the manuals?

Not everyone would appear to agree though with the underlying idea that each company should develop their own, ship specific, safety management systems — which is one of the core ideas of the ISM Code. The managing director of a Greek ship owning company expressed his view in the following terms:

"The ISM Code would be better succeed its proposal if a model manual with minimum requirements adopted by all parties concerned placed into force. The current status allows any party concerned to make own rules."

Of course the ISM Code, in paragraphs 4 and 5 of the preamble, makes it very clear that ship operators are encouraged to steer away from model manuals and to make their own rules — at least in the sense of developing their own system. Whilst the views of the Greek MD might be totally contrary to the clear and stated intentions of the ISM Code itself, it is suggested that we should not discard his thoughts out of hand. There appears to be a presupposition behind the ISM Code that every ship operating company and every ship will be able to develop a SMS in their own unique way and make it work for them. We should perhaps always be prepared to challenge any such presuppositions and consider whether that ideal situation is capable of being achieved in every instance. If it is not, is the only solution to force that company and those ships out of existence? Such thoughts should not be considered in any way as being defeatist, merely keeping an open mind and being prepared to look at all problems in a healthy and constructive way.

It is not just the masters and seafarers who have expressed concern about the dangers and utter folly of off the shelf systems. A Port State control inspector shared the following observations:

"Many companies (I feel) just go through the motions, they must be seen to do the right thing. In many cases the ISM system is bought from somewhere else with no consideration to the specific ship in question. The result is a nightmare of papers and procedures that no person is able to use as it should have been. The crews are complaining about the mass of papers that bears no relevance to their daily work situation. Ashore they are probably happy, they have a system and certificates to prove it. I think that there is a difference between the ones audited by the classification societies and the flag authorities, in my opinion the latter is the better."

Clearly, in light of the *Eurasian Dream* decision any ship operator who does have such a system in place is steaming straight into very dangerous waters. The comments of the judge clearly indicate that an inadequate/inappropriate set of procedures manuals may very well render a vessel unseaworthy. As such the ship operator may lose any right to rely upon the Hague-Visby defences, for example. There must also be a risk that the ship operator may lose its right to limit its financial liability and may even lose its insurance cover. The full consequences and implications of an inadequate SMS should not be underestimated.

In the *Eurasian Dream* judgement, which is considered in some detail in Chapter 5, the judge created quite a long list of activities, all of which would no doubt be classed as key shipboard operations as anticipated by Section 7 of the Code and no doubt for another incident an equally long list could be produced. Clearly there is a danger of going over the top and trying to provide a detailed procedure for every possible and conceivable situation. That surely cannot be the intention since the number of volumes of procedures manuals would quickly be back up to double figures. A balance is needed to avoid the dangers highlighted in the following comments received from a Port State control inspector:

"The theory of ISM is excellent. Most masters and chief engineers I interview express a frustration with the volume of record keeping they are required to maintain, while at the same time they state that ISM has definitely helped develop a more safety conscious culture which is good for the crews, vessels and environment. The two worst ISM systems are: 1) In which there is such a plethora of detail that it overwhelms the vessel's crew, thereby creating problems and 2) In which the system is so generalised that it provides too little guidance for the crew and is virtually worthless. The best systems provide enough guidance yet do not snow under the crews with avalanches of forms. I regularly inspect vessels which use the ISM forms but do so by rote, checking off items without really determining that the items have been properly addressed."

The importance of involving those on board, in particular, should never be under estimated. The full sense of ownership of the system will develop as a natural consequence from direct involvement and active participation in all aspects. An Indian second engineer put forward a very useful four part formula which should help lead towards that feeling of ownership of the system and the development of a true safety culture:

- *"Ship staff involvement should be given high priority for making company policies.*
- *Top managers on ship should be able to write each other's reports in form of performance.*
- *Transparency of work on ship and office is of prime importance.*
- *Blaming attitudes should be avoided."*

It is very likely that some, maybe many, ship operators will have to look very carefully at the SMS documentation systems they currently have in place and recognise that perhaps they need to undertake a major overhaul. In some cases it may be a matter of starting again from scratch. It is strongly recommended that ship operators look at the development of their SMS as an investment; an opportunity to make their ships and their operation more efficient as well as safe. With that efficiency and increased level of safety will come a reduction in accidents, claims and other uninsured losses. It will be an exercise to plug the drains and stop all that money being allowed to flow away. However, to do it properly requires commitment from the top of the organisation and belief that it can work. Those who are going to be directly involved in the implementation of the SMS should be consulted and be involved in the design and construction stage.

The greatest waste of money, time and other resources is to try and plough on with an inadequate/inappropriate SMS in the hope that nothing serious will happen and that there is sufficient in place to pull the wool over the eyes of the Port State control inspectors — they are becoming more sophisticated and knowledgeable as each day goes by!

The Designated Person

The full significance of the role of the Designated Person (DP) is still far from clear. This was a new position created by the ISM Code although, apparently, the idea was not incorporated until fairly late in the drafting process of the Code. There is still much speculation about the role and legal exposure of the DP amongst lawyers and academics but, as far as the author is aware, there have been no judicial decisions providing clarification of the areas of doubt. The author is also unaware of any prosecutions against a DP by the English courts.

The role of the Designated Person

The Code defines the role of the DP in section 4, where it states:

> ### 4 Designated person(s)
>
> To ensure the safe operation of each ship and to provide a link between the Company and those on board, every company, as appropriate, should designate a person ashore having direct access to the highest level of management. The responsibility and authority of the designated person or persons should include monitoring

131

the safety and pollution-prevention aspects of the operation of each ship and ensuring that adequate resources and shore-based support are applied, as required.

Interestingly, there is no real suggestion or guidelines as to the actual qualifications or experience of the DP within the ISM Code itself. As the following surveyor points out — there does not appear, on the face of it, any specific requirements as to the DP's qualifications:

"There are no provisions on the ISM nor on the SOLAS and neither on the STCW 95 as regards to qualifications to be held by the Designated Person Ashore" (Spanish Surveyor)

The ISM Code was incorporated into English Law through Statutory Instrument *SI 1998 No 1561* – at Section 8 it provides some further guidance on the Designated Person where it states:

Designated Person

8. (1) The Company shall designate a person who shall be responsible for monitoring the safe and efficient operation of each ship with particular regard to the safety and pollution prevention aspects.

   (2) In particular, the designated person shall —

   (a) take such steps as are necessary to ensure compliance with the Company safety management system on the basis of which the Document of Compliance was issued; and

   (b) ensure that proper provision is made for each ship to be so manned, equipped and maintained that it is fit to operate in accordance with the safety management system and with statutory requirements.

   (3) The Company should ensure that the designated person —

   (a) is provided with sufficient authority and resources; and

   (b) has appropriate knowledge and sufficient experience of the operation of ships at sea and in port, to enable him to comply with (1) and (2) above.

The ICS / ISF provide some further practical suggestions on the role of the DP (ICS / ISF Guidelines):

"... The task of implementing and maintaining the SMS is a line management responsibility. Verification and monitoring activities should be carried out by a person independent of the responsibility for implementation.

The designated person(s) should be suitably qualified and experienced in the safety and pollution control aspects of ship operations and should be fully conversant with the company's safety and environmental protection policies.

The designated person(s) should have the independence and authority to report deficiencies observed to the highest level of management.

The designated person(s) should have the responsibility for organising safety audits, and should monitor that corrective action has been taken ..."

A very important role, in the view of the author and one which does not always appear to be fully appreciated, is as a crucial link between the office ashore and those on board ship (and vice versa). Whilst in many companies superintendents still perform an important function as a link between ship and shore, in many other companies that function appears to be eroding away and in some cases almost non-existent. It is argued that a good safety culture can only stem from a good company culture. A good company culture can only arise when there is a genuine feeling by everyone, ashore and on board, that they are all singing from the same song sheet. To achieve that necessary culture it is crucial that there is a good and effective system of communication — particularly where issues of safety management are concerned. A good DP is instrumental in building that bridge.

Potential legal issues involving the role of the DP

Whilst there do not appear to have been any prosecutions against a DP in the English courts a few examples have arisen in other parts of the world. Possibly the most widely reported was the incident involving *Freja Jutlandic*. It appears that criminal prosecutions were commenced by the US federal prosecutors against the owners, operators, master and DP of the vessel. The main allegation against the DP seemed to be that he had instructed the master to conceal from the US Coast Guard some temporary and potentially hazardous hull repairs which had been carried out. It was further alleged that the DP had conspired with the master to falsify log books and avoid expenses to maintain a safe and seaworthy vessel. The US prosecutor alleged that there had been oily water discharges from the vessel which occurred during a number of port calls in the United States. The penalty facing the DP, if convicted, was up to five years imprisonment and a criminal fine of up to US$250,000 under each charge.

The prosecution appears to have been frustrated for a number of reasons, partly because the owners and operators declared themselves bankrupt in Denmark. However, the principle had been well and truly established in the US that the DP is clearly identifiable and does have specific responsibilities for safety and pollution prevention. The DP is, therefore, exposed to being personally cited in both civil and criminal proceedings.

Who is the DP?

The survey tried to identify who had ended up assuming the role of DP. The list of choices was obviously not exhaustive but appears to have been adequate to catch almost all situations. The question also tested whether seafarers really did know the identity of their own DP. The result indicated that only 6% of seafarers did not know the identity of their DP, which is very encouraging.

In addition to asking the ship operator category the same question, the opportunity was also taken to establish whether the respondent was indeed the DP for their Company. It could certainly be anticipated that a questionnaire relating specifically to ISM implementation would tend to gravitate towards the desk of the DP and would tend to be of more interest to that individual. However, it was pleasing to see that the completed questionnaires were returned from almost equal numbers of DPs and other members of management and staff from within the operator's office.

Figure 8.9 — Identity of Designated Person — according to seafarers

Yes 56%   No 44%

Figure 8.10 — Designated Person — from ship operator respondents

Interestingly, there was very close agreement on the identity of the DP between the two groups. For both seafarers and ship operators there were 36% of cases where the DP was also the safety officer. There were 35% of instances from seafarers and 33% from ship operators where the DP was also an operations manager, technical manager or superintendent, i.e. line managers who probably have the task of implementing and maintaining the SMS. There were also a number of DPs in very senior positions including actual ship owner themselves as well as MD/CEO.

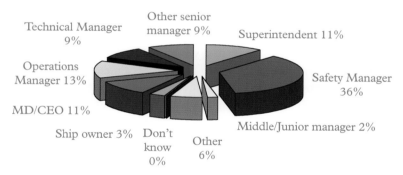

Figure 8.11 — Identity of Designated Person — ship operators

Clearly the size of the Company is going to have a bearing on the identity of the DP and what other functions that DP may perform. In a large ship management company, for example, there may well be a number of dedicated DPs who do not have any other function or maybe incorporate safety officer. In a small two ship company where there might not be more than three people in total in the office then it is unlikely, in practice, that they can afford the luxury of a dedicated DP who would not have many other job functions. In an ideal world the DP should be free to devote his or her full time to monitoring the SMS and taking what steps might be needed to ensure that it is working efficiently.

135

Potential difficulties for the DP

When the ICS/ISF suggested that "The task of implementing and maintaining the SMS is a line management responsibility. Verification and monitoring activities should be carried out by a person independent of the responsibility for implementation" they no doubt had in mind a potential conflict which might arise in certain companies between the technical and commercial operation of the ship and safety issues. This is hinted at by the following safety manager:

"The position of DP should be revised to be outside the line management of the Company, and be responsible direct to the highest levels of management (CEO/ COO). Interference from any senior company management ashore, not interested in the SMS, reduces the effectiveness of the system." (British Safety Manager)"

It is not difficult to imagine examples of situations that might arise. It may be that the ship has a very tight sailing schedule to maintain or an imminent cancelling date on the next charter to meet. The ships' staff are not properly rested, the cargo compartments need cleaning and the vessel will need to proceed at full speed through some busy shipping lanes to reach the next port on time. Whether perceived or real, the operations manager may apply pressure to sail, regardless, to meet the deadline. It may be that problems had been encountered with say the emergency fire pump which needed some new parts fitted as well as a complete overhaul. The chief engineer had submitted a report and request to the office but the superintendent responds with advice that it will be included on the work schedule for the next dry-dock which is due the following year.

In these situations the DP should have the authority and be provided with the resources, to ensure that the safety considerations are given priority over commercial or other budgetary factors. Having said that, it was encouraging to note that the real intention and potential significance was being recognised by a number of enlightened managers — such as the following operations manager who was also the DP:

"... ISM has different meaning for different people. We use ISM in an enhanced form to manage our entire operation, i.e. it is the way we work and relates to all aspects of operations, not only safety and the protection of the environment. For us it is successful. Many companies, however, require only the certification and do not actively use their SMS to enhance their management. This attitude will change with time and therefore in the longer view the ISM Code will contribute to a better managed and more professional ship management industry. Many operate at a higher level already but ISM will drag the base level higher. In summary we are better off with ISM than without it ..."

As time progresses and the full benefits of a properly implemented SMS are better understood and accepted then we should see less conflict/tension between the role of the DP and the activities of the line managers.

Internal Audits

The real purpose and meaning behind the internal audit was explained very well in the ICS/ISF Guide when it described the situation in the following terms:

"In carrying out internal SMS audits companies measure the effectiveness of their own systems. Internal audits are potentially more important than external audits for controlling the effectiveness of the system, since companies stand to gain or lose more than the external audit bodies if the system fails. The Company, its employees, shipmasters, officers and crews own the safety management system and have direct interest in ensuring that it is effective. As a result, the internal SMS audit, which represents these interests, should be at least equal to if not exceed the thoroughness of the external audit process."

Indeed a Canadian ship operator put it very strongly. He said:

"From my point of view the attitude, competence and perceived credibility of the company's internal auditors are critical to the whole process. They make it or break it!"

The requirements for undertaking internal audits are set out in Section 12 of the Code. The audit is a crucial part of the cycle of continual improvement to check whether those who are involved in implementing the SMS are actually doing what they say they are doing.

COMPANY VERIFICATION, REVIEW AND EVALUATION

12.1 The Company should carry out internal safety audits to verify whether safety and pollution-prevention activities comply with the SMS.

12.2 The Company should periodically evaluate the efficiency of and, when needed, review the SMS in accordance with procedures established by the Company.

12.3 The audits and possible corrective actions should be carried out in accordance with documented procedures.

12.4 Personnel carrying out audits should be independent of the areas being audited unless this is impracticable due to the size and the nature of the Company.

> 12.5 The results of the audits and reviews should be brought to the attention of all personnel having responsibility in the area involved.
>
> 12.6 The management personnel responsible for the area involved should take timely corrective action on deficiencies found.

Something that is of crucial importance in understanding the function of the internal audit is that the self-regulatory principles of the ISM Code makes the role of the Company paramount. Part IV of the ICS/ISF Guide sets out some 'Internal Safety Management System Audit guidelines'. What is not at all clear from Section 12 of the Code is who exactly should be conducting the internal audits on board ship. Should it be the seafarers working on board ship or should it be a superintendent or similar from the office ashore?

One thing it does say is that '*each audit is to be carried out by personnel who, at the time of the audit, are independent of the area, office or shipboard department or activity being audited*' (Section 12.4). The ICS/ISF Guide suggests that for the purpose of carrying out shipboard SMS audits, companies may find internal auditors from the following sources:

- Company managers, including safety, operations and technical managers.
- Masters, chief engineers and senior officers.
- Third party SMS auditors.

It is suggested that the background of the auditor, from the above three options, can make an enormous amount of difference to the way in which the whole of the SMS functions. Before exploring the reasons why that might be, the results of the survey are considered to see who in practice is conducting internal auditing.

The idea behind company verification, review and evaluation, in particular internal audits, is to provide a means by which the Company can measure the effectiveness of its own systems. It has been suggested from the early days of ISM implementation (see, for example, the ISF/ICS Guidelines Page 34) that internal SMS audits are potentially more important than external audits for controlling the effectiveness of the system. The important point to recognise is that it is the companies who stand to gain or lose more than the external audit bodies if the system fails. The suggestion, therefore, is that the

Company, its employees, masters, officers and crew should develop and recognise an ownership of the safety management system. In this way they will understand and appreciate that they have a direct interest in ensuring that it is effective.

Who conducts internal audits?

This question proved to be a little more difficult to answer than was first anticipated when the questionnaire was originally drafted. A significant number of both seafarers and shore-based staff appear to have found the idea of the seafarers actually conducting their own internal audits to be quite strange. Indeed it transpired from the survey that most internal audits are at least controlled, if not actually conducted, by staff from the office ashore.

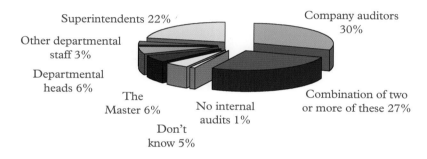

Figure 8.12 — Who conducts internal audits? — seafarers

The responses from the ship operators shore based staff mirrored the seafarers account of the situation very closely.

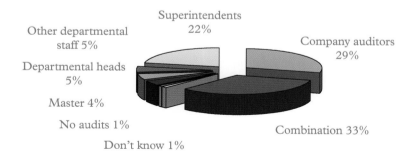

Figure 8.13 — Who conducts internal audits? — ship operators

139

The author admits to being quite surprised that the internal auditing was so closely controlled from the office ashore rather than allowing the master and those on board to conduct this most important function.

An interesting observation was made by Captain Mike Shuker in a paper he presented at the INMarEST Conference in May 2002 when he drew attention to a list of issues which he suggests are not adequately provided for in the ISM Code itself:

- *Who should carry out these internal audits?*
- *The frequency of the internal audits.*
- *Who is responsible for overseeing the internal audits?*
- *Who allocates the non conformities raised during internal audits for corrective action?*
- *Who verifies the corrective action plan?*
- *How corrective actions are verified.*

No doubt some would argue that the ISM Code expects each company to decide itself how such matters are dealt with in that particular company and that these items will be dealt with quite differently by different companies. In practice there appear to be some serious concerns being expressed about the way that auditing techniques are being developed. A chief officer put it quite simply in the following terms:

"The paper mountain generated — where will it stop! During audits, auditors seem only interested in correctly completed forms and certificates, very rarely checking any of the physical systems."

In addition to the company's own internal auditors there may be many more third party auditors attending on board. They may be representing Flag State Administration/recognised organisation, Port State control, charterers and many other categories. The operations manager from one shipping company seems to hold some very strong views about these auditors and the ISM Code:

"...The ISM Code has only created a new industry or profession of ISM auditors who are, more often than not, arrogant. They stride into the owners or managers office once a year for annual audits, asking incompetent questions to persons who have been working for decades in the profession ..."

Certainly the auditor should not adopt an arrogant attitude and it is unfortunate that this ship operator had encountered such a bad experience. An auditor who is doing his/her job properly will be acting as a facilitator and helper to the Company and the findings used in a positive way within the cycle of continual improvement.

A no blame culture

The idea of a no blame culture is seen by many as a crucial factor that needs to be developed if the fruits of the ISM system are to reach full maturity. It is amongst the most difficult of all aspects of ISM implementation to achieve. In some ways it seems to run contrary to our natural human instincts and the culture in which we have been brought up. If we do something wrong we expect to be punished. As children at home, or at school perhaps, if we kicked our football through a glass window we could predict, with a fair degree of certainty, what the consequences would be if we were caught. If no one had seen the incident and there was nothing to link us to the crime and if we thought we could get away with it we might have looked the other way and pretended we knew nothing about it. It is a matter of self preservation. A no blame culture takes us beyond the punishment factor although that does not mean diminishing the responsibility factor. We would still be expected to pay for the replacement window (or at least make a contribution towards it). The no blame culture would anticipate the incident being used as a learning opportunity. Why did the window get broken? What could be done to prevent a recurrence? What would be the consequences of making changes to avoid a recurrence — i.e. it forms part of a risk management exercise. If not provided with that learning opportunity then we would continue to have broken windows.

Most people have been so conditioned by the blame culture they really have difficulty even believing that anything else could possibly work. In addition, even if they did move into such a culture could they be certain that their superiors, who perhaps had control of their careers and jobs in their hands, were also fully committed to that no blame culture. These fears were echoed by Stuart Witherington of the MAIB in a paper he presented at the IMarEST Conference in May 2002 when he said:

"...society's blame culture instils into managers and seafarers a fear of blame and criminalisation. It encourages mistrust, preventing them from being open and honest, by covering up mistakes when things have gone wrong. Further, it can give a sense of anxiety to individuals who think that by taking personal responsibility, they may be held responsible for an accident simply by following the dictates of the ISM Code ..."

Of the limited number of safety management systems which the author has seen which are working very well, they are invariably a long way down the road of having developed a no blame culture. From the author's own first hand experience a no blame culture is quite rare and takes a long time to cultivate. It was quite a surprise therefore to review the findings of the survey when a majority of respondents, both seafarers and ship operators office staff said that their companies really did operate a no blame culture.

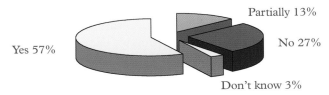

Figure 8.14 — Does the company operate a no blame culture? — seafarer's perception

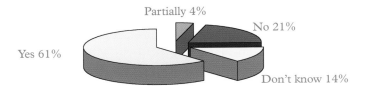

Figure 8.15 — Do you operate a no blame culture? — ship operators' perception

In fact, the results from this question and the next relating to the safety culture were very difficult to reconcile with other answers provided, for example, in the sections dealing with reporting accidents, hazardous occurrences and near misses as well as many of the narrative comments generally.

The questionnaire not only asked ship operators whether they believed that they operated a no blame culture but also asked them to try and predict how their seafarers might answer that question. In other words, do seafarers believe that the Company really does operate a no blame culture. Figure 8.16 shows the ship operators response.

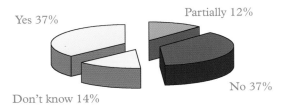

Figure 8.16 — Does the Company operate a no blame culture? — ship operators' prediction of seafarers responses

142

The scepticism of the ship operators appears to have been a little over-pessimistic. This can be seen by comparing the actual responses from the seafarers alongside the perceptions of the ship operators.

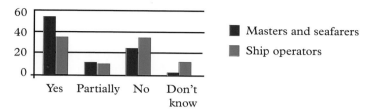

Figure 8.17 — Does the company operate a no blame culture? — a comparison of views

These differences of perceptions between what the seafarers seem to think and what their managers and superintendents ashore believe they think arose on a number of occasions during this study. The predictability of the differences of perceptions probably points towards a problem with communications. The next chapter re-examines whether seafarers really are working in a no blame culture.

A safety culture

Sitting side-by-side with a no blame culture is the idea of a safety culture. These could be thought of as the two supports for the SMS. If one support is not stable and strong, or otherwise crumbles, then the SMS cannot stand and will also crumble.

In his address at the 25th annual World Maritime Day on 26th September 2002, the Secretary General of IMO also stressed the importance of this issue in the title of his paper: 'Safer shipping demands a safety culture' (IMO Briefing 26 Sept. 2002). Mr. O'Neil drew attention to the fact that rules and regulations are not in themselves sufficient where safety and environmental protection are concerned.

> *"Although the behaviour of individuals may be influenced by a set of rules,"* he said, *"it is their attitude to the rules that really determines the culture. Do they comply because they want to, or because they have to? To be truly effective in achieving the goal of safer shipping, it is important that the shipping community as a whole should develop a want-to attitude."*

It is difficult trying to explain in words what is actually meant by a safety culture. Perhaps the most concise description known to the author is: 'the raising of safety to the highest priority.' An interesting definition has been provided by Professor Jim Reason (possibly quoting a source from the CBI) who suggests that it is '*the way we do things around here*' (Reason p.173). The Health and Safety Executive of the UK provide a somewhat more technical definition: '*Safety culture of an organisation is a product of individual and group values, attitudes, competencies and patterns of behaviour that determine style and proficiency of the organisations safety programs*'. What this means in practice is making sure, on every occasion, that the safety issues are considered and appropriate steps taken before undertaking any task or operation. If, having conducted such a risk assessment the conclusion is that the task cannot be undertaken safely in the existing circumstances then it is not allowed to take place. Combined with the no blame culture, which would remove the pressure when taking such decisions, the probability of accidents occurring will be reduced considerably. Just think about how many accidents are caused because we cut corners or took risks because we believed that getting the job done was more important than getting the job done safely. Often the reasons are because of actual or perceived pressure/threats from our superiors. For example, a ship is operating on a tight sailing schedule. The master and the mates have not had any rest for 36 hours but the ship must sail to maintain its schedule. An assessment of the situation would quickly identify that the master and mates are tired/fatigued and it would pose a significant risk to allow them to navigate the ship to its next port. Within a safety and no blame culture, options would be explored, but if the risk was high then the decision to delay the sailing until the master and mates had been rested would be taken without hesitation and without fear of repercussions. Some ship operators and indeed seafarers, would find such a suggestion quite unrealistic and quite outrageous. The answer to them is surely that if you think suffering the losses involved in failing to maintain a schedule will be expensive, try having an accident!

Without doubt, the development of a true safety culture, like the no blame culture, will require a lot of time, effort and hard work to achieve. Many long established and inbuilt prejudices will need to be overcome — it really will involve a major culture shift. It was a further surprise to see that again a significant majority of both seafarers and ship operators seem to believe that they do work within a safety culture.

These answers do not seem to sit at all comfortably with other answers given elsewhere in the questionnaire and the narrative comments received. It is suspected that the respondents, in answering

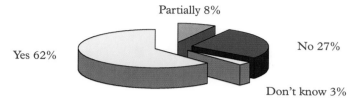

Figure 8.18 — Do you work within a safety culture? — seafarers

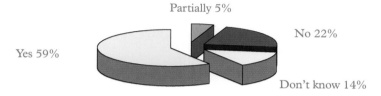

Figure 8.19 — Do you work within a safety culture? — ship operators

these questions, were perhaps indicating that, in their view, they worked in a reasonably safe manner. But does that really extend to feeling confident that the ship's schedule can be interrupted if it was felt necessary to allow the crew to be rested? If it does then full implementation of ISM is probably closer than the author might suggest is the case in a later chapter.

A similar exercise was carried out by asking ship operators about the safety culture, as it was with the no blame culture. That is they were asked not only for their own views but also how they thought the seafarers would answer that question. Again the ship operators were perhaps a little cautious and sceptical about the seafarers perceptions and views about a safety culture;

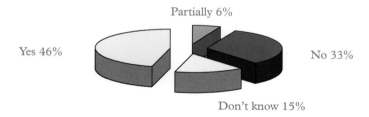

Figure 8.20 — Do seafarers believe you operate a safety culture?

To appreciate these perceptions in perspective it is useful to compare them side by side:

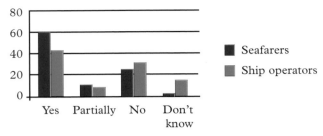

Figure 8.21 — Safety culture — a comparison of views

The IMO have recently been highlighting the importance of developing similar ideas:

In order to promote a no-blame culture the suggestions of the committee to member governments was to:

- *Review their regulatory and safety regime with a view to encouraging the reporting of near misses without fear of reprisal or punitive action.*

- *Urge companies operating ships under their flag not to penalise persons reporting near misses.*

- *Urge companies operating ships under their flags to implement procedures by which persons should only report near misses to the designated person or persons and the designated person or persons should only pass on such reports in an anonymous form.*

[IMO MSC/Circ.1015]

Some of these issues relating to the development and implementation of no-blame and safety cultures are considered further in subsequent chapters.

Ownership of the SMS

Another factor that many believe to be crucial to the success of an SMS is the idea that those most closely involved in its implementation should develop a sense of ownership of the system. This is a very strong motivational factor.

If a ready-made system is presented to the ship then there will be an immediate risk of engendering a sense of alienation, rejection and probable resentment. It will probably be perceived as yet another set of rules and regulations, along with a whole lot of paperwork, which is being imposed on those on board. On the other hand, if the system is devised with the active participation of those on board then it will engender a sense of identity and purpose. Those who will be involved in the implementation process will have had some say in its development. It will be their system and they will, as a natural consequence, want to make it work. It has perhaps not yet occurred to some, but the people who will benefit most from a properly implemented SMS will be the seafarers themselves.

When asked whether they felt a sense of ownership of the system the two categories of respondents were actually quite positive:

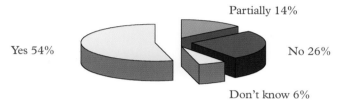

Figure 8.22 — Is there ownership of the SMS? — seafarers

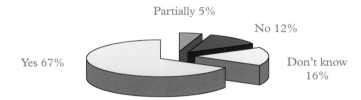

Figure 8.23 — Is there ownership of the SMS? — ship operators

These responses also do not sit comfortably with the narrative comments received and other answers given elsewhere in the questionnaires. However, on the face of it, the perceptions of the seafarers and the staff ashore seem to coincide quite closely as can be seen in the comparison below:

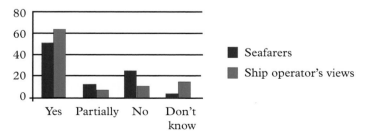

Figure 8.24 — Is there ownership of the SMS? — a comparison of views

A sense of ownership of the safety management system can only come from active participation and involvement in the system. Where there is such active participation then the results are clear to see. It really should come as no great surprise to find that when sea staff themselves are actively involved in the development of the SMS then it stands a very good chance of achieving success. The converse is almost certainly true. On the positive side there were some excellent illustrations submitted of how that sense of ownership was bearing fruit. It could not be expressed more clearly than the following received from an Australian master:

> *"The SMS within the Company was developed by seafarers, for seafarers. The vessels feel a great sense of ownership because the procedures and work instructions, checklists etc. were all developed by the ship's staff. This is indeed an enlightened approach and gives the ship's officers a sense of ownership and relevance."*

A great sense of pride and achievement can be gained on the road to developing that sense of ownership. For most companies there was actually little that was new with the ISM Code. They already had good safe procedures in operation which had been developed over many years. The main difference was that now those good safe procedures would be written down in such a way that everyone in the Company could follow the same procedures in a consistent way. It is not difficult to empathise with those who prepared their procedures manuals in this way when they reflected and realised that they had been involved in developing and running some pretty good procedures and that now they were going to have the opportunity of passing on their knowledge, skills and experience to others through the safety manuals. A little of that sense of pride and achievement can be detected in the following report from a US master:

> *"Our safe management plan was written by seasoned mariners who worked for this Company by collecting processes from ships. Every possible process via educating officers to use flow diagrams — this information was consolidated and became our safe management system. I was one of those who contributed."*

Not only must the old hands feel that sense of pride and ownership — it must involve staff at all levels both on board ship and ashore and in every type of organisation. We may naturally think in terms of bulk carriers or tankers where there is a greater propensity for the relatively small team who will be together perhaps for six months or more at a time on relatively long sea passages with relatively few port calls to cooperate and work at developing the safety culture and the sense of ownership of the system. What about other craft though — maybe a fast ferry operation — involving very frequent port calls with relatively very short sea passages between ports? Can an SMS really work on board such craft? Can any sense of ownership of the system be developed? Clearly it can and one Danish master made the point very clearly:

> *"The following has been my own way to succeed with the implementation and maintenance of ISM in a fast ferry company in frequent service carrying more than 1·5m pax per annum. It is essential that the ISM system is designed and implemented in cooperation with all groups of ship/shore personnel and management. The staff must end up with a clear feeling and pride that this is their system. They made it...*
>
> *Reporting procedures are the backbones of the system. Internal audits performed by inter-department staff must be scheduled for the full year and the schedule must be kept. The audit report must be published in full for everyone to learn from others mistakes. Policies and procedures must be reviewed in planned intervals and brush-up training in system management should be performed on regular basis."*

Whatever the ship type, whatever the operation might be, the feedback received clearly demonstrates the need for the development of that sense of ownership of the system and a clear understanding of how that sense of ownership is achieved. Additional questions relating to the significance of ownership are also considered further in the following chapters.

Recruitment policy

To complete a review of this section of the questionnaire, ship operators were asked one further question, relating to their policy on recruitment. This was asked so as to allow the author to consider whether there might be any connection between recruitment policy and attitudes towards ISM. The results show that the majority of ship operator respondents were involved more in employing their staff directly or through their own captive manning agents rather than using external recruiting agents (see figure 8.25 overleaf).

During the latter part of the 1990s and into the new millennium many companies recognised the mistake that had been made in the past and realised that they again needed to start developing those vitally important bonds between employer and employee. The way in which manning agents are used in some companies can inhibit the possibility of such bonding taking place. The ship operators recognised that there most definitely was a correlation between the employer/employee relationship and a well run/safe ship. Unfortunately, others

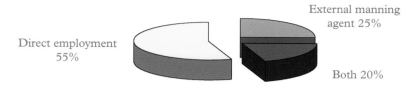

External manning
agent 25%

Direct employment
55%

Both 20%

Figure 8.25 — Recruitment policy

are taking a little longer to realise this very important, fundamental, fact. Others recognise it but need persuading of its significance. A Mexican chief officer explained his experience in the following way:

"Training on ISM is an important matter sometimes neglected by the Company as most of the time they worried about the evidence that the person went throughout the familiarisation on the subject, an easy exam and that's forever .

— No further interest on effectiveness of the system is checked or evaluated, only those which can be audited or required by the certified body (class).

— Keeping a permanent or continuous sea staff is also difficult as the Company has officers and ratings going and coming and in this way you lose continuity on the purpose of maintaining a safety culture.

— It is important that everyone considers the importance of the ISM Code and SMS even when the economic part plays a roll on the decision process."

Employers themselves also recognise the importance of continuity of employment and some of the implications and significance was explained by the following Swedish operations manager:

"High turnover prevent SMS to work efficient. Training in basics is ever ongoing and never completed because it has to start all over again. Crews are too small to enable training on SMS, occupational work and do the work properly at the same time. Crew from manning agents not always taking company goals and objectives to their heart because they will be gone after a few trips. Lack of ownership because short employment. Others with long term employment recognise however the importance of a functional SMS because it makes their job easier by providing routines and the basis to train the newcomers."

Not only is the continuity of employment issue important in itself — to provide a bond between employer and employee — it is also vitally important from a pragmatic and very practical reason, within the context of ISM, because people then have an opportunity of becoming familiar with one working system. Through that familiarisation they stand a real chance of making the SMS work, which would mean, at the end of the day, that ships really would get safer and seas really would become cleaner!

150

It is almost inconceivable to think of any other industry where the owner of a plant, worth many millions of dollars, would hand over the management to individuals who were engaged on a casual labour basis but that seems to be exactly what some people in the shipping industry did and some continue to do. The ISM Code is, in many ways, a reaction to those unbelievable practices.

Chapter 9

THE SURVEY RESULTS ON REPORTING ACCIDENTS, HAZARDOUS OCCURRENCES AND NON-CONFORMITIES

Introduction

The reporting of accidents has been quite a normal occurrence for many years and for certain accidents it has been mandatory to report. The Flag State Administration or the Port State may have a requirement and they may have specific forms that need to be completed and particular aspects that they require recording in the accident report — particularly if personal injuries are involved or pollution has occurred.

Under the UK Regulations, masters and skippers of fishing boats must ensure that the circumstances of every accident and serious injury are examined. Under the Merchant Shipping and Fishing Vessel (Health and Safety at Work) Regulations 1997 (SI 1997 No. 2962) the safety officer should conduct an investigation and the results must be sent to the Chief Inspector of Marine Accidents at the Marine Accident Investigation Branch (MAIB). Reports of accidents must be reported within 24 hours by telephone, fax, telex or e-mail, and reports of serious injuries must be sent to the MAIB within 14 days, using the quickest means possible.

The classification society may also require certain accidents to be reported — particularly if the hull of the ship or its machinery has been damaged. The ship operators are likely to require accident reports — particularly if the accident is likely to lead to a claim either being made against them or a claim that they may need to make against their own insurers. Many of the accidents would be analysed although often the extent of the analysis and investigation would only go so far as establishing where legal liability might rest.

That said however, there have been a number of ship operators who have looked for causes of accidents as part of a process of trying to learn from those events and implementing some form of corrective action to prevent the same event happening again. The ISM Code does devote an entire section to reporting not only accidents but also hazardous occurrences and non-conformities. The requirements are set out in Section 9:

9.1 The SMS should include procedures ensuring that non-conformities, accidents and hazardous situations are reported to the Company, investigated and analysed with the objective of improving safety and pollution prevention.

9.2 The Company should establish procedures for the implementation of corrective action.

For the requirements of Section 9 to be complied with fully the author believes that the concepts of safety culture and no blame culture, discussed in the last chapter, must have been developed to a fairly advanced stage. It is also the view of the author that if Section 9 is being complied with properly then there is a very good chance that the rest of the SMS will also be working well. Of course the proactive side of the SMS, utilising risk assessment tools, is very important. It is when the point is reached where near misses can comfortably and confidently be reported, analysed and steps taken to implement corrective action that the SMS is clearly alive and dynamic. In 2001 IMO published a detailed paper under MSC/Circ.1015 'Reporting near misses' — which highlighted the importance of this level of reporting.

It was Section 9 of the Code, therefore, which was to be the focus of the survey and which the author believes will provide a clear and reasonably objective insight into the current status of ISM implementation. The survey not only attempted to establish which types of incident were reported and how many, but also the extent to which these incidents were analysed and any lessons learnt fed back into the system — within a cycle of continual improvement. It also attempted to establish whether there might be any reasons why people might be reluctant to report incidents. If there was, then clearly such inhibitions could act as a brake to advancing the development of safety on board.

An Indian engineer put forward a very positive picture:

"ISM Code has drastically reduced the number of accidents, raised the standard of ships. Near miss reports are a valuable guideline to the ships personnel. ISM Code can eliminate the substandard ships. It will ensure safety of personnel, ships and cleaner seas."

It should also be recognised though that there is a very real possibility that both companies and individuals might not actually understand

153

the basic principles of reporting incidents other than actual accidents. A classification society auditor touched upon this important issue:

"Although most of the companies are doing their best to improve their safety and pollution prevention systems, people on board are not fully familiarised with reporting NCN, accidents, etc. maybe because they don't have a clear idea of how to, and they feel that it represents extra admin work, which if not properly managed could give them more problems.

I have also noted that there is confusion for both internal auditors and persons on board in distinguishing between a NCN and a technical deficiency.

The most serious problem that I have seen is the lack of understanding of the real meaning of corrective action normally because investigation and analysis of NCN is not adequately carried out, the instinct being to choose an immediate solution to avoid spending further time on the subject."

An Indian pumpman made a very important point which is worth thinking about, for many could perhaps learn an important lesson about investigating and analysing incidents:

"The reports of accidents on board should be made by enquiring about it from everybody on board and not only the heads of the departments. According to me this will bring out the proper picture of the accidents."

Which incidents are reported?

The questionnaire adopted a position whereby it treated the incidents to be reported as falling into the three broad categories as stated in the heading of Section 9 of the ISM Code, i.e. accidents, hazardous occurrences and non-conformities. In an attempt to obtain an impression of the general attitude towards reporting, the questionnaire asked the respondents to state, initially, what level of incident was reported for the three different categories. The respondents were given four options to identify the level of reporting:

- Every incident was reported.
- Most incidents were reported.
- Only serious incidents were reported.
- None, i.e. no reporting took place.

The respondents could also choose a don't know option. Some respondents chose not to identify any preference. For some individual respondents their answers may be very precise and to that extent objective. For example, the master or safety officer completing the questionnaire should know exactly what they report. If some other member of the crew then their opinion may be based on what they have been told by someone else or otherwise be a perception, although given in good faith and thus be subjective in nature. It is very important to realise that the responses provided by the masters and seafarers and the ship operator categories should contain reasonable first hand answers or at least be close to the factual situation. The other stakeholder category is almost certainly going to be expressing an opinion as to what they think the situation is. There is nothing wrong with that of course, since this category has the potential to adopt an impartial observer status, but it is necessary to keep that in mind when reviewing the graphs and considering the findings. The graphs below show the responses from the three different respondent groups to the three categories of incidents and are colour coded in accordance with the established convention.

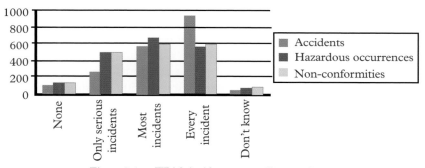

Figure 9.1 —Which incidents reported? — seafarers

Although every accident should be reported it would appear that the seafarers are being honest and recognise that perhaps not every accident that occurs really is reported. However, the vast majority claim that, if not every accident, then at least most accidents are reported. It is of concern though that over 250 seafarers indicated that only serious accidents were reported and 87 stated that accidents were not reported. Interestingly a very significant number of seafarers indicated that every or most hazardous occurrences and non-conformities were being reported. This is very important and its significance will be considered presently.

The ship operators seemed less convinced that every accident was being reported but were satisfied that most were. They were even less in agreement that every hazardous occurrence and non-conformity was being reported when compared with the seafarers submissions.

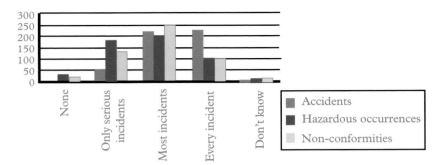

Figure 9.2 — Which incidents reported? — ship operators

Whilst they indicated that most hazardous occurrences and non-conformities were being reported they were starting to tend towards only the serious incidents end of the scale.

The other stakeholder category produced a very interesting perspective. It is important to remember that what is being reported here is the individual's perception of the situation rather than being based on any factual evidence. Very few respondents in this category believe that every accident is reported and only about 50% of the remainder believe that even most of the accidents are reported. Few seem to believe that there is any significant reporting of either hazardous occurrences or non-conformities. Understandably a larger number of this category registered that they did not know.

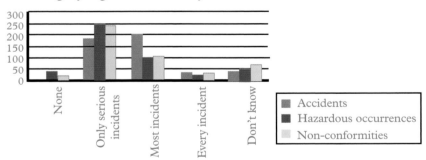

Figure 9.3 — Which incidents reported? — other stakeholders

It is when these findings are compared side-by-side, and expressed in percentage terms, that something very significant starts to appear:

The responses from the seafarers and the ship operator groups were not too far apart in their perception of which incidents were being reported but the other stakeholder category had formed a very different view altogether. Why there might be such big differences in

156

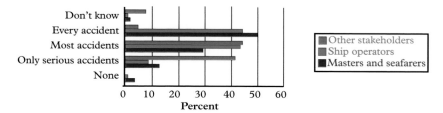

Figure 9.4 —Which accidents are reported?

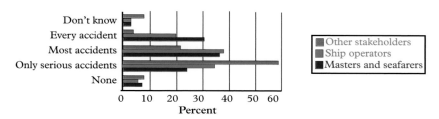

Figure 9.5—Which hazardous occurrences are reported?

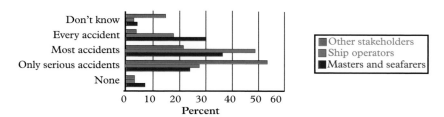

Figure 9.6—Which non-conformities are reported?

perception will perhaps start to become clear as we progress through this analytical process.

Before we consider how many incidents are reported it is perhaps worth reflecting upon an observation submitted by a Filipino rating on which incidents are reported:

> *"Most officers report unpleasant incidents, occurrences or near misses if it's the crew or ratings fault but indeed they are directly responsible.*
>
> *If it's their own fault they just make up for it or put it in writing in case port authorities come on board.*
>
> *In my own observations and experiences some vessels (including my present vessel) has already implemented ISM and with valid SMC but they don't follow it religiously or honestly. Instead they just file it and make it ready for inspectors or port authorities. It's only for formalities sake."*

157

How many incidents are reported?

The next question in the questionnaire asked the respondent to indicate how many of the different types of incidents were reported each year per ship. Again the degree of objectivity within the answer would clearly depend upon the actual knowledge of the individual respondent. No doubt the master, safety officer and DP ashore would have a very accurate figure available whereas others might be guessing, although on the basis of their experiences on board.

Because of the way the database had been set up it was not possible to allow respondents to choose a specific number. Rather they were given a choice from a range of options as follows:

- Zero
- 1—10
- 10—25
- 25—50
- More than 50
- Don't know

There was a considerable degree of agreement between all three groups and therefore individual group analysis results have not been reproduced here. Rather the comparative results, in percentage terms, are shown for each category of incident.

Approximately 80% of all respondents seemed to be in agreement that between 1 and 10 accidents, hazardous occurrences and non-conformities were reported from each ship each year. There was a slight increase in the numbers of hazardous occurrences and non-conformities reported.

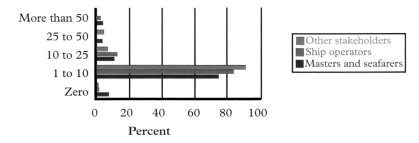

Figure 9.7 — How many accidents are reported?

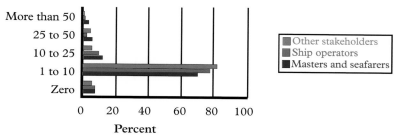

Figure 9.8 — How many hazardous occurrences are reported?

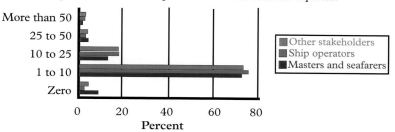

Figure 9.9 — How many non-conformities are reported?

When we compare these results we can see that the numbers of accidents, hazardous occurrences and non-conformities were almost all falling into the 1—10 reported incidents per ship per year.

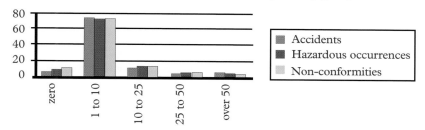

Figure 9.10 — How many incidents? — masters and seafarers

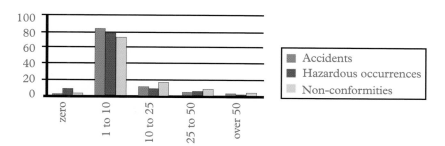

Figure 9.11 — How many incidents? — ship operators

159

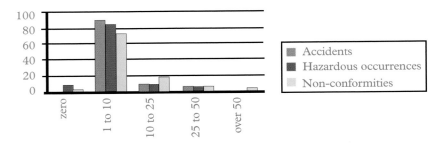

Figure 9.12 — How many incidents? — other stakeholders

The indications are therefore, at least according to the masters and seafarers and ship operators that most/all accidents, hazardous occurrences and non-conformities are being reported and that there are between one and 10 of each type of incident being reported each year per ship. This is perhaps a good opportunity of considering the theory behind the Accident Pyramid.

The accident pyramid

It has been recognised for many years that there seems to be a clear relationship by way of a ratio between the number of major incidents, the number of minor incidents and the number of near misses. Indeed it is this principle that is applied when, for example, the terms of insurance on a car are proposed. Usually there will be an excess or franchise, typically the first £100, which the insured will bear before the insurance company becomes involved. The reason is that the car insurance company will have conducted a risk assessment and concluded that the majority of accidents tend to be small in nature, i.e. below £100 and they would prefer not to cover those smaller, frequent, accidents. The larger the excess the individual is prepared to accept, the lower will be the premium. The reduction in premium though is not at all linear. The curve is closer to being exponential in shape, reflecting the general idea of the probability of serious accidents compared to smaller type incidents.

This idea has frequently been illustrated as an isosceles triangle or pyramid shape. There are a number of versions of the so called accident pyramid or safety pyramid — a typical example is shown in figure 9.13.

If this pyramid or triangle theory is correct then we should expect to see significantly more hazardous occurrences and non-conformities than actual accidents. The survey results do not appear to support the

160

Figure 9.13 — A tpyical accident pyramid

theory — the numbers of accidents, hazardous occurrences and non-conformities all appeared to be equal. Even allowing for the fact that the numbers involved were a little vague, i.e. somewhere between one and 10, there still seems to be something not quite right. There are at least two possible explanations:

1   The pyramid theory is wrong.

2   The other stakeholders were right and the seafarers and ship operators were not reporting all hazardous occurrences and non-conformities.

Whilst it may be the subject of some debate and disagreement as to exactly how steep the slope might be on any particular accident pyramid, few would challenge the general principle. A clear understanding of the implications of the accident pyramid must be understood if the full significance of reporting hazardous occurrences and near misses is to be fully appreciated.

It is suggested that it is infinitely more cost effective and efficient to encourage near miss reporting and transcend the current blame culture than to wait for accidents to happen. It is perhaps worth exploring, therefore, a little more of what is involved in near miss reporting.

Causal chains

There are a number of theories about causation. The traditional view is that a causal chain can be identified whereby one event is the cause which leads to an effect and that effect then becomes the cause of another effect and so it goes on, cause — effect, cause — effect. The author believes this is too simplistic and a somewhat misleading explanation. Rather, any major incident is the result of a whole series of multiple causal factors — small chains — all coming together at one particular geographical location at a particular point in time. The interruption of any one of these multi-causal events would prevent the particular incident happening when or where it did. It may be that the occurrence of the major incident had only been postponed and therefore it is crucial to ensure that the remedial steps taken are sufficiently significant in nature to reduce to a minimum the probability or possibility of the incident becoming an inevitability.

Hazardous occurrences or near misses, which constitute the base of the accident pyramid, occur much more frequently than more serious accidents. They are also smaller in scale, relatively simpler to analyse and easier to resolve. Each major accident can usually be linked to a number of incidents that happened earlier. By addressing these precursors effectively, large and expensive accidents may be avoided.

The proposition is that if it were possible to examine and deal with the large number of incidents, or near incidents towards the base of the pyramid, then the probability of occurrence of the serious incidents at the peak would be reduced considerably. The suggestion is that serious/major accidents do not just happen in isolation — they have causes which can be identified. Frequently, following a major incident, there will be an inquiry of one description or another, depending just how serious the incident was. With the benefit of hindsight it is usually possible for the investigators to identify a whole series of causal factors, frequently developing over a lengthy period of time, that eventually led to the major incident occurring.

It has been difficult obtaining details of cost involved in the formal investigations. The National Audit Office were receptive, kind and helpful but did not have any figures. Through the offices of the MAIB it was possible to establish that the cost to the British government (or British taxpayer depending upon how you may wish to look at it) of say the *Derbyshire* investigations up to and including the reopening of the formal investigation are estimated as totalling between nine and ten million pounds, including legal fees. The *Bowbelle* and *Marchioness* incident, including the current formal investigation, is estimated at some six million pounds, including legal fees.

The author suggests that we do not need to wait for these major incidents to occur. Invariably, warning signs would have been flashing up in advance of the incident actually happening. These warning signs are causal factors starting to influence events. All the causal factors have not yet come together at the same point in time in the same location. The causal factors manifest themselves as hazardous occurrences, near misses, non-conformities, etc. We can explore and examine these causal factors by investigating and analysing them. By identifying and dealing with them, by applying corrective actions, they are removed from the equation and the major incident will be avoided. It is suggested that this has surely got to be an infinitely better method of dealing with incidents than waiting for the major disasters to occur, with substantial loss of life or major pollution. That is why it is so important that near misses are identified for what they are and taken very seriously.

A near miss is an event that signals a system weakness that, if not remedied, could lead to significant consequences in the future. As such, a near miss is also an opportunity — an opportunity to improve system structure and stability and an opportunity to reduce risk exposure to potential catastrophe. (Wharton)

If this theory is correct, therefore, it is possible to identify and deal with the causal factors before the major incident occurs. It is the hazardous occurrences, near-misses and non-conformities which are providing the warnings and opportunities to implement the necessary corrective action before lives are lost, pollution occurs or other losses are incurred.

Learning opportunities

The concept and reason for reporting not only accidents but also near misses and hazardous situations, as is required by the Code, is very clearly explained and brought to life by an Irish master who reported:

"All accidents/near misses/dangerous occurrences are reported so that all the other vessels in the fleet can maybe benefit from our experiences and visa versa. The majority of non-conformities can be dealt with on board quite soon and even these are promulgated throughout the fleet."

Another good example of the system working as it was intended was reported by a Finnish master who has clearly moved far away from the blame culture and would now appear to live in a true safety culture. He describes the system in his company in the following terms:

"In our ship, all accidents and incidents are reported. This is a standard procedure here and we have had no problems with that. We are receiving monthly, 'lessons learned' reports from accidents/incidents, happened onboard our ships. These are also followed up very closely to avoid any further acc/inc. in the future."

To a significant degree ISM is about transparency — transparency at a number of different levels including reporting of hazardous situations and near misses. To achieve the level of transparency required it is necessary to address the issues in the very way in which the SMS itself is structured and also in developing the confidence of personnel so that they can operate without undue fear of reprisals in an open and transparent system. A British master reflected on the sort of general feeling that is needed to start moving forward when he stated:

"I consider our system has promoted a more open approach to reporting incidents and the investigation and corrective action arising from such reports."

There is probably a natural reluctance on the part of human beings to report something if there is a risk that somehow that same individual is perhaps going to be punished, blamed or otherwise criticised for the particular event. If the cause of accidents is to be fully understood and if accidents are to be prevented before they happen then it is crucial that near misses and hazardous situations are reported, analysed and dealt with. In different parts of the industry a variety of incentive schemes have been introduced to try and encourage compliance with the ISM Code or otherwise to develop a safety culture.

Whether it be an internal SMS audit, or some other safety management audit by the ships own staff or some external body such as a Port State control inspector or Flag State/recognised organisation (usually a classification society) auditor the opportunity will present itself for detecting non-conformities. The detection and recording of a non-conformity should not necessarily be considered as something bad or viewed in a negative way since it is basically providing a learning opportunity. For one reason or another the non-conformity is flashing a warning light that the system is not working as it was intended, or at least as it was defined or described in the written procedures. It may be that someone had misunderstood or misinterpreted the procedure, in which case some further training/familiarisation might be required. It may be that the procedure itself was flawed in some way, in which case the corrective action may be to amend the procedure. It may be that an individual was deliberately refusing to comply and follow the procedure, in which case the individual would need to decide whether they would comply or otherwise they would have to face the consequences, clearly depending on the severity of the non-compliance. It can be anticipated that even within the best structured SMS some problems will manifest themselves as non-conformities. That is fine provided procedures are in place to pick up those non-conformities, analyse them and determine why they arose and implement corrective action. In this way the SMS is the subject of a cycle of continual improvement. By a process of fine tuning, the system will be getting better and better as time goes on.

Of course, there will be varying degrees of seriousness applied to non-conformities. What are being discussed here can be described as minor non-conformities although they may indicate the development of a potentially serious situation if remedial action is not taken. A much more serious and very different situation arises when a major non-conformity is identified. A definition is provided which will perhaps help distinguish this occurrence from the minor category event:

- Major non-conformity means "an identifiable deviation which poses a serious threat to personnel or ship safety or a serious risk to the environment and requires immediate corrective action." In addition, the lack of effective and systematic implementation of a requirement of the ISM Code is also considered as a major non-conformity.

For present purposes the focus is on minor non-conformities. Within the context of the present discussion the important point to recognise is that non-conformities should be capable of being discovered through an audit process. It may be possible to try and hide or disguise such non-conformities, but an experienced auditor would probably discover irregularities which would certainly raise his suspicions.

It may be that some of these non-conformities do, in themselves, represent hazardous situations. However, through the audit process those hazardous situations would be discovered and dealt with. It is suggested that the hazardous situations referred to in Section 9.1 of the ISM Code are events different from non-conformities.

Although the ISM Code itself does not mention near misses this expression does throw some light on the type of event contemplated by the term hazardous situation. It is the accident which nearly happened. If things had been just slightly different — a few seconds in time or a few centimetres in distance — there would almost certainly have been an injury or some other damage or loss. A narrow escape we might say, or a close shave, is another expression which might be used. One or two examples might help to explain not only the nature of the hazardous situation event but also the very real problem associated with reporting such events:

Example 1

A seaman was assigned to paint some bulkheads on the outside of the accommodation. He couldn't reach the upper part of the bulkhead and so he found a ladder which he leant against the bulkhead. As he started to climb the ladder the foot of the ladder slipped and he jumped off the second rung. He was not injured but a little paint was spilt.

Example 2

A vessel is on passage between the third and fourth loading port. It is 0230 hours and the second mate is alone on the bridge as the officer of the watch. The second mate had only managed to catch four hours sleep during the last two days because of rapid turnround in the load ports. He must have fallen asleep because when he looked up he saw another vessel crossing on his starboard bow about half a mile away. He immediately altered course and managed to avoid a collision.

In example 1, whilst the seaman might be somewhat embarrassed at being careless and not following correct safety procedures, he would probably recognise the sense in telling the bosun or mate what had happened and ensuring that a second man was assigned to stand by the bottom of the ladder and/or ensure the ladder was otherwise secured before starting to climb or indeed whether it was safe to undertake the job. Maybe the ship was rolling excessively. Whilst it may seem somewhat extreme to consider preparing a formal report for such an incident, the point is that this could very easily have resulted in a very serious accident or even a fatality. If that seaman could make the mistake then others might also make a similar mistake. Better that the lesson be shared with as many others as possible to remind them to follow the correct procedures. That dissemination of information would naturally follow a report and analysis. That, in simple terms, is the logic and philosophy behind near miss reporting.

Now, consider example 2. Is the second mate going to advise the master of what has happened? If he did, how will the master react? If the master was advised and recognised that the second mate was suffering from fatigue, that he should not have been left in charge of a watch and certainly should not have been alone on the bridge, is the master going to advise the DP in the office ashore? If he did, how would the DP react?

166

If put in the position of the second mate and asked that question it is suggested that the reaction of many in the shipping industry, both shore based and seagoing — if they were honest — would probably be that they would keep quiet about the incident if they thought they could keep it a secret. Clearly what is being advocated and indeed required by the ISM Code is very much the opposite. The reason is not difficult to understand.

If the second mate had not woken up when he did, had perhaps slept for just two minutes longer and if the second mate on the other ship was also not aware of the situation which was developing the hazardous situation would possibly become a major accident. Lives might be lost with personal injuries, pollution, explosions, serious damage to both vessels, maybe vessels sinking and damage to cargo with consequent loss of time and earning capacity. In that case the facts would most likely come out in a formal inquiry and/or the subsequent investigations in anticipation of litigation or insurance claims. The fact that the second mate was seriously fatigued, that he was asleep and alone on the bridge, would not only seriously prejudice the ship operator's legal position but also have implications as far as the licence of the second mate and the master are concerned.

If the second mate and possibly master and whatever other mates might be on board are working such long hours that they are suffering from fatigue at this level and there are not sufficient seamen to ensure that an additional lookout is maintained then that ship is not safe — indeed it is unseaworthy. There is a very serious problem on board which needs addressing. The near miss was a warning signal. Such an incident must be reported immediately to the DPA and immediate steps taken to remedy the situation. It is suggested that such remedial steps would not include dismissing or otherwise disciplining the master and second mate but rather provide them with the additional resources necessary to operate the ship safely. In other words, if it was intended to maintain a very tight loading schedule then put on board additional qualified mates sufficient to ensure that excessive hours were not worked by any one individual and consequently that fatigue was not allowed to take hold. Also, additional seamen may have to be provided on board to provide adequate lookouts. Maybe some ship operators would argue that such luxuries cannot be afforded in the current depressed market conditions. The point is to consider the cost of the alternative!

The situation, though, is probably that the culture of fear in which many seafarers seem to work would discourage them from reporting such an incident and take a risk that they might also get away with it a second time. The IMO has recognised the problem of the reluctance to report near misses and has issued guidelines on developing a no blame culture as previously discussed.

Unfortunately there were a significant number of respondents to the survey who did not appear to have grasped the full potential value of accident or near miss reporting as a tool to prevent future incidents. Indeed, it would appear that the concept of a safety culture may still be a little way off. For example, the following passenger ferry master seems to have resigned himself to the fact that accidents are inevitable when you carry large numbers of people and only very serious incidents could warrant the time needed to report:

"The figure of 50+ accidents must be viewed in the context of number of souls carried which on average is 3,900 every day. Obviously we don't expect to have corrective action reports for every minor sprain, burn and cut."

Of course, sympathy must be expressed at the potential dilemma of such a master who does not have unlimited resources available. However, for so long as the attitude is maintained that these minor accidents will continue to happen there is a very good chance that they will! Once the cause of these incidents is investigated and analysed then there should be a very real chance that something can be done to prevent further incidents in the future. Admittedly, it involves additional work initially, but once the problem is addressed and corrective action put in place then the actual work involved reduces to a maintenance programme.

Another example of the nature of misunderstandings with regard to the purpose of near miss reporting can be seen from the following report from a British master:

"Not all agreed on what constitutes a near miss e.g. chipping hammer head coming loose and dropping off its shaft was promulgated as a near miss!"

Presumably, from the way the master has expressed his comment, he thinks such an incident is not a near miss. He seems to be suggesting that those who think otherwise have perhaps made a mockery of the idea. The author would suggest that such an incident is a very good example of a near miss. The heads of chipping hammers should not come off! Presumably, on the occasion referred to by the master, no one was injured, but it could very easily have been otherwise. Surely, if the head of one chipping hammer has dropped off it would make sense, good seamanship sense, to check the other chipping hammers to make sure none of the others are loose, to make sure that any defective hammers are repaired and made safe and to remind the bosun, or whoever has the responsibility for such equipment, to make sure that they are regularly checked and that everyone who uses them are reminded to check them. Is that not just good seamanship, the way we had always worked? All ISM is doing is putting a little more structure into that good practice. If anyone was still tempted to say that we do not need such a formalised system then surely the facts of this case speak for themselves. The informal system had, for whatever reasons, failed. The chipping hammer had not been maintained properly, such that the head had been allowed to get to such a state that it dropped off. Fortunately, on that occasion no-one was injured, but a very important lesson can be learnt.

Another observation was received from an Australian second engineer who also was not convinced that near miss reporting could serve any useful purpose. He had the following to say:

"I work in the engine room compartment where you use hand tools (drills, grinders etc), machine tools (drill press, lathe etc), welding machines. The vessel moves, therefore we must move as well as carry out our work whilst using machinery. This in itself is dangerous — you can reduce the risk — i.e. correct PPE — ensure balanced — job held firm but still it is a hazardous work environment at all times. So to fill out near misses is pointless."

Of course the engine room of a ship is, potentially, a very dangerous environment and when the ship is at sea the risk of accidents happening is increased even further. Hazards will always be there and proactive steps must be taken to reduce the risk of those hazards developing into actual injuries, pollution or damage incidents. However, an analysis of almost all the many accidents that do still occur in many engine rooms on board ships, every year, will prove that the incidents should not have happened. They could have been prevented if certain steps had been taken.

Since there are so many actual injuries and other incidents every year it is suggested that there will be many more hazardous situations and near misses. If these are looked for, identified, reported, analysed and corrective action taken then it will help considerably to reduce the number of actual injuries and losses. The fact that someone works in a potentially hazardous environment should make them even more acutely aware of the risk of accidents happening and they should be looking even more closely for learning opportunities to reduce the risk. Reporting hazardous situations and near misses provides an excellent opportunity to do this in a structured way and to pass on such experiences to others who might also benefit.

Yet another example of this misunderstanding which seems to exist in some quarters with regard to the nature and value of reporting and the potential for learning lessons was submitted by a British master:

> *"A lot of the Corrective Action Reports (CARs) that stem from internal Company ISM audits are petty and serve no purpose to anyone. CARs are raised against ratings who remain completely baffled by the whole exercise."*

Surely if the ratings remain baffled by the whole exercise then something has gone very seriously wrong with the SMS in that company, in particular with communications, training and familiarisation on board that ship. The master and the Company he works for really do need to look very carefully at the way their SMS has been set up and the provision of training and familiarisation. In particular they should look at how the SMS works, the crew's involvement in its operation and the importance of reporting as a cycle of continuous improvement. They may well find, as other masters and companies have done already, that by providing leadership and involving the crew in the SMS they can make a very valuable and positive contribution.

The study from the Wharton School Risk Management Centre suggests that, apart from safety improvement through the identification and resolution of isolated near misses, there are additional safety and management benefits of a near-miss program. They say in their Phase One report of December 2000 that these include:

1   Delegation of Safety Responsibility: An effective near miss program shifts the task of identifying unsafe operations from Environmental, Health and Safety (EHS) management, to a much larger work force that has intimate contact with process operations/equipment. By harnessing this larger work force a greater number of safety related issues can be identified and addressed.

2   Increased Safety Awareness: By making individuals more safety conscious and by shifting the responsibility of identification of near misses, unsafe conditions and behaviour to each individual in the work force, both on and off the job — safety of employees can be improved significantly.

3   Creation of an Information Pool: The collection and analysis of near miss data can reduce accident frequency through a) identification of similar incident precursors at other facilities, and b) pattern observation and trend analysis over time. Such a

knowledge base would reduce risk exposure in ongoing operations as well as future equipment, process and plant design.

A fundamental question arises in this consideration of near miss reporting. Is the identification of a large number of near misses indicative of a safe or unsafe on board operation? Could it validly be claimed that the identification of a large number of near misses suggests that there are serious problems and the whole system is unsafe? Alternatively, could it be argued that because there have been many near misses identified that the seafarers have become more safety conscious. The point is that good safety management actively looks for near misses and accidents are resolved proactively before they occur.

If there is doubt about what the correct answer might be, consider the following three hypothetical scenarios and consider which ship is most probably operating an efficient, functioning SMS:

1   Ship A appears to be operating close to perfection. It has submitted no reports of any accidents, hazardous occurrences, near misses or other non-conformities for the whole year.

2   Ship B has submitted two separate accident reports. One was in respect of an injury to a seaman who slipped on some spilt oil, broke his leg and was sent ashore to hospital. The second report related to a pilot who fell off the pilot ladder, seriously injuring his back. There were no other reports of any other accidents, hazardous occurrences, near-misses or other non-conformities.

3   Ship C has submitted reports in respect of one major incident, five quite serious incidents, 21 minor incidents and 53 hazardous occurrences and near-misses.

It is certainly possible for a ship to operate without experiencing major, serious incidents and possibly to reduce minor accidents to small numbers but it is very unlikely indeed, so long as human beings are involved in the operation of ships, that there would be no hazardous occurrences or near-misses of one description or other during the year. The results from Ship A, therefore, should be viewed with considerable suspicion, probably suggesting that there is no effective reporting procedure on board that ship at all and thus a malfunctioning or non-existent SMS.

It would be very difficult for a ship not to report such serious accidents as those which resulted in a seaman being hospitalised and a pilot falling off a ladder. In addition to any mandatory requirement to report such incidents, they are also very likely to result in claims against the ship owner who in turn would look to his P&I Club for the appropriate insurance coverage. However, it is very difficult to imagine that if a ship had experienced two serious accidents, as in the case of Ship B, that these would have occurred in total isolation. It is much more likely that such incidents were indicative of inadequate considerations of safety issues and, consequently, one would expect to find many smaller incidents, hazardous occurrences and near misses. Since none were reported one would be led towards a conclusion that reporting is limited to an absolute minimum and that there is probably little or no effective safety management on board.

Is Ship C a floating disaster area? With all those incidents and near misses surely there is a serious problem on board this ship with regard to managing safety? Not so — Ship C is demonstrating a typical pattern of a vessel where the safety management system is starting to work. Provided all the incidents are the subject of proper corrective actions, the SMS can then be fine tuned to prevent future recurrence. It can then be expected that the number of accidents, hazardous occurrences and near misses will reduce over time.

It became apparent from a number of survey responses that reports are being sent in from vessels of accidents, hazardous situations and non-conformities but with no feedback from the office. Clearly nothing could be more demotivating and demoralising. A British master described the way things work in his company:

"As far as the ISM Code affects my operation the whole thing is a complete 'paper chase' and is seen as such by most of us. We send in non-conformities and hazardous reports and invariably they are not answered, it's all a question of cost. The audits are regularly done but most of the time its only going through the motions."

The audit system should of course be working both ways and in such a case there should be serious non-conformities raised against the office ashore and the DP. A similar alarming situation was reported by a chief engineer:

We work for 'An Oil Major' (the name was supplied) and our working period is 90 days on 45 off — I feel that more safety aspects should be looked at as to who the person is who gives some of these lead audit people their certificates and try and make them into gods — which they think they are. At present time I should say 985 of our company employees are sick of having ISM Code rammed down our throats and nothing is done when we make reports. All our safety audit people do is cover their own backsides when it suits them.

Clearly, if that chief engineer was typical of the other 984 employees referred to then it is highly unlikely that there could be any semblance of a working SMS in place and it may well be in order to raise non-conformities against the auditors, the Company and the DP.

For others there is clearly an understanding of what the ISM Code is trying to achieve, but they are having difficulties balancing all the different pressures upon them. A typical example is the following comment from a chief officer:

"There are a number of minor accidents and some near misses that go unreported. Many of these occur during the busiest times on a vessel and often the paperwork involved will take a very low priority over the pile of paperwork for operational and other requirements. In my opinion, although ISM is a good thing and has been embraced practically by my company it will always suffer from a extra paperwork stigma being attached to it."

Clearly the dilemma is one of managing time and priorities. If the incident was not too serious and the drafting of the report can wait until the ship is at sea then that is clearly a reasonable thing to do. However, if the incident was serious, or potentially serious, then commercial pressures may have to come in second place. It is going to take some time before the shipping industry accepts that approach. Many readers may well have found themselves sitting in an aeroplane on the runway ready to take off, only to be advised that a problem has been identified and the plane is going nowhere until the problem is solved or the potential problem removed. Whilst it may be felt to be something of an inconvenience, few passengers on the plane would seriously suggest that the pilot should ignore the warning light, or whatever the problem might be, take a risk and fly the plane regardless! The question which arises is why, in the shipping industry are we prepared to tolerate taking a ship to sea when there is a known or perceived problem? Can commercial pressure ever be such that we are prepared to take those risks? No doubt many would say we have always taken those risks in the shipping industry and that is part of our culture. Perhaps those same people should reflect a little though and consider to what extent those attitudes actually led to the need for the ISM Code being developed in the first place. Perhaps some of the readers will recall the facts surrounding the *Herald of Free Enterprise* disaster in 1987?

A cycle of continual improvement

Within Section 9 of the ISM Code is a concept of continually improving the SMS and the general management of safety on board by learning lessons from accidents or events which nearly became accidents. Of course all would agree that it would be infinitely preferable to prevent accidents happening in the first place, but if they do occur — or something serious nearly happens — then these events should be recognised as learning opportunities. They are opportunities to fine tune the SMS such that corrective actions can be implemented to reduce to a minimum the possibility of such an event arising again. By this process of fine tuning and checking that the corrective action has worked, the SMS will lead progressively to a safer and safer ship operation. This concept can be illustrated in a simple flow diagram:

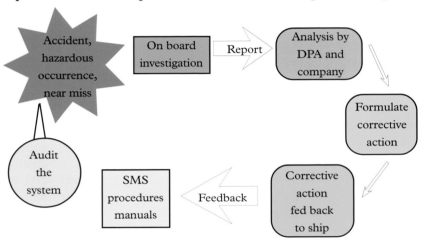

Figure 9.14 — A cycle of continual improvement

The ISM Code allows each individual company and ship to develop systems which best suits their own requirements and the way they do things. It is not possible, therefore, to lay down one reporting system that would apply to every single company and ship in the world. However, the rather simple flow diagram in figure 9.14 illustrates the basic processes which should be followed.

If an incident occurs on board — it doesn't need to be an actual accident — it could be a hazardous occurrence or near miss or some other warning bell ringing or a non-conformity that has been identified. It should then be noted and reported on board. Clearly the way things develop from there depends on the actual nature and circumstances of the particular incident. If it is a major incident then professional lawyers and investigators will probably arrive on board very quickly to assist the master with the investigation and collection of evidence.

174

However, if we assume for the purpose of this explanation that the incident was relatively minor, we can follow the various steps of the process around the cycle of continual improvement.

1   The incident occurs and the most appropriate steps are taken to deal with the incident.

2   The incident is reported/noted.

3   An onboard investigation is undertaken to establish the causal factors.

4   The causal factors which led to the incident are analysed.

5   An immediate, temporary, corrective action plan is instigated.

6   A report of the incident is submitted to the Company, probably the DP.

7   The report is analysed by the Company.

8   Feedback to the ship is provided by way of a corrective action report. This may confirm or amend the suggested corrective action and may involve amending the formal procedures.

9   If appropriate, advice is circulated to the rest of the fleet in order that as many as possible can take advantage of the learning opportunity and ensure that there is not a recurrence.

10   The new procedure is formally implemented and communicated to all concerned.

11   After an agreed period of time, the relevant part of the system is audited to ensure that the corrective action has had the desired effect.

The survey questionnaire went on to establish to what extent the cycle was being completed. Following reports being submitted to the Company, were corrective action reports being sent back to the ships and was an audit process in place to check that the corrective actions were having the desired effect?

Corrective action reports

From the responses received from both the seafarers and even more so from the ship operators, the majority of reports submitted in respect of accidents, hazardous occurrences and non-conformities do appear to generate corrective action reports, i.e. feedback from the office to the ship. An analysis of the seafarers responses is shown in the graphs in figures 9.15, 9.16 and 9.17:

Figure 9.15 — Are corrective action reports returned? — accidents — seafarers

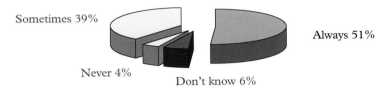

Figure 9.16 — Are corrective action reports given? — hazardous occurrences

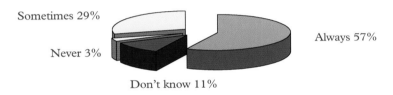

Figure 9.17 — Are corrective action reports given? — non-conformities

Both the seafarers and ship operator groups reported unequivocally that corrective action reports are always returned although approximately one third suggested the softer sometimes. The positive significance of this can perhaps be seen more clearly when we group the three types of incidents together — as in figures 9.18 and 9.19.

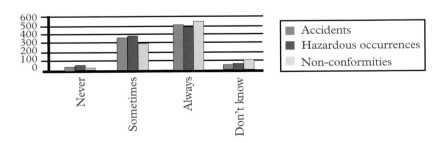

Figure 9.18 — Are corrective action reports given? — seafarers

176

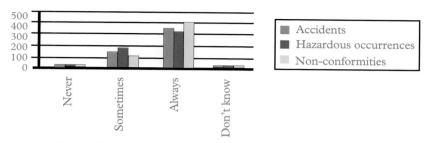

Figure 9.19 — Are corrective action reports given? — ship operators

However, the other stakeholder group was again in serious disagreement with the other two groups of respondents. The perception of the other stakeholder group seemed to be that only a quarter believed that corrective action reports were always returned. They were prepared to give some benefit of the doubt though and allow for the possibility that corrective action reports might be returned sometimes.

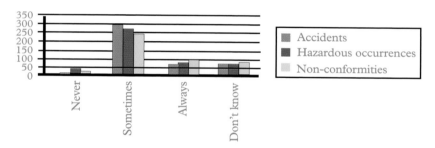

Figure 9.20 — Are corrective action reports given? — other stakeholders

If the seafarers and ship operators are being accurate in their recording that most reported incidents are the subject of a corrective action report, then the ISM implementation process would appear to be at a much more advanced stage than had first appeared to be the case.

An important aspect of corrective actions is that they can be considered as the lessons to be learnt from a particular incident. Certainly, feedback must be provided to the ship where the incident actually occurred, but the lesson to be learnt should be shared around the fleet and even further if possible. This will then allow maximum advantage to be taken from an unfortunate incident. However, some caution does need to be taken as was pointed out by a Ghanaian chief officer:

"The ISM Code is being implemented by crews/officers from varied backgrounds. Hence internal auditors should not expect same occurrences on sister ships. A corrective action on one ship should only serve as a guideline when attempting to bring it to the notice of personnel on a sister ship."

As with everything else related to ISM, common sense must always prevail. Everyone must keep their brains in gear. ISM does not expect people to stop thinking and become some type of robot. One of the aims is to introduce consistency but if something is clearly not applicable or relevant to a particular vessel then it is difficult to understand how it could possibly contribute to improving safety. If there is a conflict then the matter should be taken up with the DP.

Follow up audits

The survey results with regard to the follow up audits were quite similar to the corrective action reports, except that there was a slightly less positive response with less respondents saying always, more saying sometimes and more saying never.

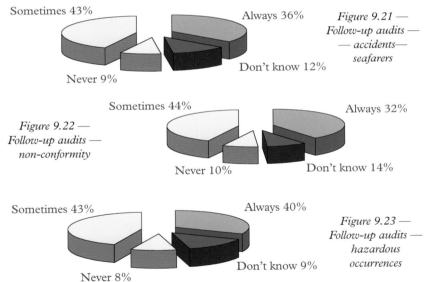

Sometimes 43%   Always 36%
Figure 9.21 —
Follow-up audits —
— accidents—
seafarers
Never 9%   Don't know 12%

Sometimes 44%   Always 32%
Figure 9.22 —
Follow-up audits —
non-conformity
Never 10%   Don't know 14%

Sometimes 43%   Always 40%
Figure 9.23 —
Follow-up audits —
hazardous
occurrences
Never 8%   Don't know 9%

When the comparative pictures are looked at we still see a very similar pattern with general confirmation from the seafarers and ship operators that about a third are conducting follow up audits on all corrective actions and another half on some of them. The other stakeholder group continue to remains sceptical.

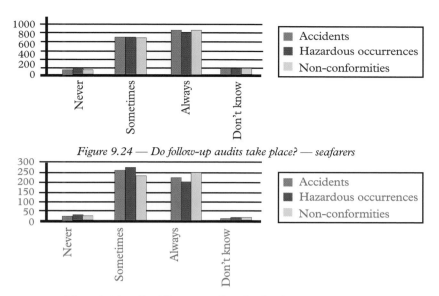

Figure 9.24 — Do follow-up audits take place? — seafarers

Figure 9.25 — Do follow-up audits take place? — ship operators

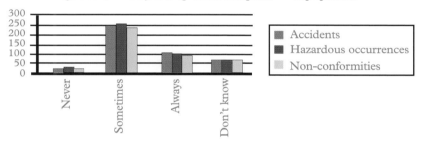

Figure 9.26 — Do follow up-audits take place? — other stakeholders

It is quite possible that the importance and full significance of conducting follow up audits, within the context of the cycle of continual improvement, is not fully appreciated by some. An Australian ISM consultant explained the importance very well:

"Since the focus of the questionnaire appears to be NCs, accidents and incidents, I can honestly say that this is one area that few companies manage very well and thus fail to reap the full benefits of the reporting system. Most have good reporting systems, but fail to follow up effectively. Firstly, there is a lack of timely corrective action and, secondly, there is usually no follow-up to see that the proposed corrective action was effective. Near miss reporting and follow-up is almost non-existent, even in the best of companies, usually due to lack of understanding and training."

On the basis of the answers given in the survey, at least as far as the master and seafarer and ship operator groups are concerned, everything

179

would appear to be proceeding very well indeed towards compliance with Section 9. Consequently, if the author is correct in his assessment of the significance of Section 9 as an indicator of general compliance, then the majority of the respondents are well on their way to full compliance with the ISM Code including the implementation of no blame safety cultures.

Perhaps it would be helpful to recap where we have got to so far, as far as master and seafarers and ship operators are concerned:

- The majority believe that the Company they work for promotes a no blame culture.
- The majority believe that they work within a safety culture.
- The majority believe that they have ownership of their SMS.
- Almost all accidents, hazardous occurrences and non-conformities are being reported.
- Corrective action reports are being returned for almost all of the reported incidents.
- Most of the corrective actions are being implemented.

Many of these results/conclusions would appear to be contradicted by the answers given to the next series of questions in the questionnaire. They also appear to be contradicted by the narrative comments included at the end of the questionnaire.

Reluctance to report

It is certainly possible that objections and criticisms could be raised against the series of questions in the questionnaire asking whether there was any reluctance to report incidents. In one sense they were perhaps leading questions, but on the other they provided an opportunity for the respondent to open up a little in his/her responses to the questionnaire. The responses provided to this series are extremely interesting and enlightening and start to provide a clearer insight into the current status of people's real attitudes towards ISM and the implementation of the SMS as well as such things as the so-called no blame and safety cultures.

The questions were almost rhetorical in nature, suggesting that if there was any reluctance to report incidents then why should that be. The questions provided ten possible reasons, plus an open box in which the respondent could insert a different reason. The respondents were invited to identify up to three factors, in order of priority, of reasons why they were reluctant to report. The respondents were given the option to state that there was no reluctance to report. That option would clearly render the question academic as far as that respondent was concerned. However, what actually happened was that most of the respondents, who had previously indicated that they were reporting everything and doing all that they were supposed to be doing, actually identified a whole range of reasons why they were reluctant to report.

Recalling the answers provided to the earlier questions about no blame and safety cultures, one could reasonably have expected that there would be no reluctance at all to reporting accidents at the least and probably little reluctance to reporting hazardous occurrences and non-conformities. That did not turn out to be the case though. Rather surprisingly, nearly 50% of the seafarer respondents said that they were reluctant to report even accidents.

The way the ship operator's version of the questionnaire was phrased meant that they were giving their view as to whether or not they thought there might be any reluctance on the part of the seafarers to report. It was even more surprising, therefore, to find that almost two thirds believed that the seafarers would be reluctant to report accidents. When it came to analysing how the other stakeholder category responded it was really quite staggering that only 10% believed that seafarers would not be reluctant to report accidents.

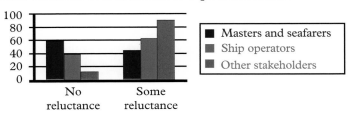

Figure 9.27 — Reluctance to report accidents — comparison of perceptions

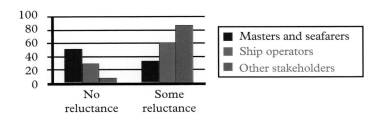

Figure 9.28 — Reluctance to report hazardous occurrences — comparison of perceptions

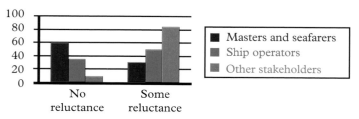

Figure 9.29 — Reluctance to report non-conformities — comparison of perceptions

Almost identical results were received with regard to reluctance to report hazardous occurrences and non-conformities.

The way the questionnaire had been structured was intended to give the respondent a choice. Either they ticked the box to say there was no reluctance to report or to tick up to three boxes to indicate why there might be a reluctance to report. What happened, though, was that many respondents ticked the box indicating that there was no reluctance to report but then went on to tick further boxes indicating why there was a reluctance to report! Clearly this was contradictory in itself but it also seemed to contradict the earlier answers in which the majority of respondents seemed quite clear that they believed their companies operated a no blame culture and they were working within a safety culture.

It is important that we keep these apparent contradictions in mind as we now explore some of the reasons why the seafarers are reluctant to report accidents as well as hazardous occurrences and non-conformities and why the ship operators and other stakeholders believe the seafarers have reluctance.

Reluctance to report — views of masters and seafarers

In answer to the earlier question, almost all seafarers said that they reported most, if not all, accidents. It really is of considerable concern therefore, that the same seafarer respondents should now declare that they are reluctant to report accidents. An analysis of the reasons given, and the scale of the concern, is shown in figure 9.30, which also includes a comparison with the reasons why there might be reluctance to report hazardous occurrences and non-conformities.

In the questionnaire the respondents who stated that they were reluctant to report were asked to list up to three reasons and to rank them in order of priority. When constructing the graph in figure 9.30 the method used for determining a figure to use was to consider a first

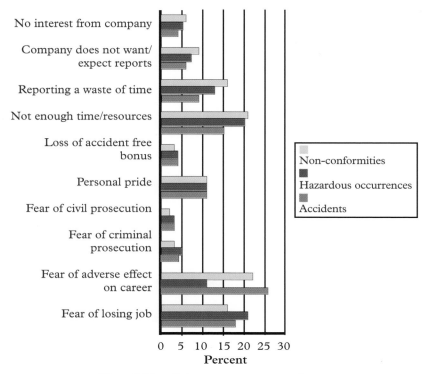

Figure 9.30 — Reluctance to report — seafarers

priority choice as three units, a second priority choice as two units and a third priority choice as one unit. In this way a total number of units were allocated to each option and the graphs drawn to show those units as a percentage of the whole.

It becomes clear from the analysis of the results that seafarers fear the repercussions that reporting might bring to their employment and career prospects. This could perhaps have been expected as far as reporting hazardous occurrences are concerned, i.e. where one might be admitting a mistake which otherwise no one else would have known about. But being frightened of reporting actual accidents is worrying. Of course, if there really was a no blame culture in existence then reporting all incidents would be seen as a rewarding experience, a learning opportunity and any fears should have been dispelled long ago. It is important, however, to analyse the figures as they are and try and make sense of them.

Certainly the fear of losing one's job or seeing career prospects damaged did show up to be a significant factor. An ISM consultant who did have first hand experience identified a particular incident of which he was aware:

"… Look at for instance chemical tankers in main ports such as Rotterdam. The officers and crew are working around the clock. This is well known to the Company, but if the crew complain they get no answer or if they don't like the Company they could find other work… It will take many, many years yet before we will get a good safety system implemented in full."

A Filipino master set out the problem and the solution very clearly:

"In my opinion, to maximise reporting of hazardous occurrences/near misses and to eliminate reluctance to report same, there must be some ways or provisions in the Code that will guarantee the person who report that he/she will have no fear of adverse effect on career prospects, losing job and any prosecution."

Another report from a chief mate proposed a similar need, this time based upon thoroughly disgraceful behaviour by his employers. He shared the following experience:

"There needs to be a guarantee of protection for the mariner who is reporting or is part of a situation being reported. I received a Letter of Warning from the USCG for not following SMS procedures over a Near Miss incident where it was discussed in a safety meeting and then a crew member anonymously passed the information to the USCG. A piece of atmospheric test equipment failed and personnel entered 10 feet into a cargo tank not fit for entry. These experienced people didn't like the smell of the tank and immediately exited with no harm. I got no support or backup from the Company when I was approached by the USCG and the only way the details of this incident would have been available to anyone other than those involved was directly through the near miss discussion. I was demoted to 2nd mate and never promoted back up to my C/M position of 10 years. I am now sailing C/M with another company with another horrible SMS."

Hopefully the chief mate will eventually find an employer who is equally enlightened and will reward that type of reporting which would very likely help to save lives.

The author had anticipated, when he first constructed the questionnaire, that if there was any reluctance to report then it would probably be because people were afraid of either civil or criminal prosecutions being commenced and the reports being used as evidence against the originators. That proved not to be the case at all. The survey results indicate that, amongst seafarers, there is little worry about such prosecutions and certainly that was not a concern that would make them reluctant to report the incidents.

The extent to which masters in particular, but also other seafarers, are exposed to criminal sanctions, fines and even lengthy jail sentences for a wide variety of offences including non-compliance with the ISM Code is really quite staggering.

A Chinese master felt that the master needed reassurance that he could rely upon the provisions of IMO Resolution A.433(XI) — Decisions of the Shipmaster with regard to Maritime Safety and Marine Environmental Protection — which was intended to give the master legal protection if he did carry out his job as he was supposed to do. He also felt that the ship operators should be exposed to criminal sanctions. His conclusion, and solution are set out below:

> *"After three years, my conclusion is ISM has not improved our marine safety and environment protection but makes worst. My suggestions:*
>
> *(1) National legislation should reinforce the spirit of IMO Resolution A.443(XI) which protects the masters from unjustifiable dismissal;*
>
> *(2) The shipowners, operators and DPs should be criminally liable for their wilful default or gross negligence, and the penalty should include imprisonment;*
>
> *(3) DP shall not be the same person as operation manager or anyone involve in the daily operating activities."*

Under UK legislation the ship operator and DP, as well as the master and anyone else involved in the operation of the SMS, may be exposed to a wide range of criminal sanctions for various offences relating to the implementation of the ISM Code.

Personal pride was recorded as a factor that made some seafarers reluctant to report incidents. This is a very human response. No one likes to admit that they fouled up, or nearly fouled up. However, it is this factor which needs to be overcome in the no blame culture such that these incidents are examined as learning opportunities. The most important issue is not to find someone to blame but rather to find out what went wrong and establish what needs to be done to prevent a recurrence.

In some companies, in an attempt to encourage the development of a safety culture, they offer their staff various bonuses or other incentives by way of a reward if they reduce the number of accidents on board. This is a good idea in principle but could, in some cases, encourage seafarers to stop reporting incidents for fear they would lose their bonuses. The accidents had not stopped — they had just been brushed under the carpet, so to speak. The fear of losing the accident free bonus was therefore included as a possible option that the seafarers could choose. In fact it turned out to be not an issue at all. Very few respondents indicated any concern with accident free bonuses. Similar schemes involve actually rewarding seafarers if they do send in reports. Such a scheme was described by the safety manager in a Greek shipping company:

"I believe that one useful tool in fighting the fear of reporting should be the adoption of an ethical reward scheme to ships that come forward with their non-conforming conditions and we should then focus on how to rectify this (possible) unsafe condition instead of focusing solely on the correctness and exactness that the reporting form has been completed."

This is possibly an excellent scheme which could work very well and achieve the desired results if managed properly. However, once again it could be open to abuse in that spurious or irrelevant reports were being submitted, purely to obtain the reward.

One of the most significant issues was the feeling that there just was not enough time to report. This was repeatedly the subject of detailed narrative comments received from a wide range of respondents. The inference was that if there was more time and perhaps other resources, then reporting would be undertaken without reluctance.

There seems to be an understanding amongst many in the industry that the SMS has to be paper based. The author has had difficulty understaning this perception since there is no reason in his mind why the SMS should not be run on an electronic system, provided it is adequately backed up and the people who need to access the system can actually do so. Indeed there are some very good software programs available on the market that allow a ship operator to construct their own electronic SMS. Such an electronic system can help reduce labour intensive paper based reporting considerably. A marine surveyor raised this very issue :

"The ISM Code has made management ashore more aware of the operational difficulties encountered on board their vessel(s), however, many masters omplain about the volume of reading material and paperwork. Many ships' officers allege the Company they work for operates no differently from before ISM was implemented. This surveyor believes the ISM Code has had a beneficial effect, but the expectations may have been greater than the actual result. What may be more useful than the volumes of ISM Code books found in the shipboard offices, would be a software package, employing computers now found on board, which would simplify the mandated reporting procedures, supplemented by an Email protocol. Most ocean-going vessels now employ this technology and it should be used more effectively in a real time reporting environment."

On a related matter, the author is aware of a very exiting project, which is at an advanced stage of development, that allows the use of portable and fixed electronic recording devices to be used to directly enter data into a PC and thus creat records and reports almost automatically. The officer must still physically undertake the checks but uses the electronic device as a labour saving tool. As of August 2003 the system is undergoing trials onboard a number of different types of ship.

Of considerable concern was a relatively significant number of masters and seafarers who stated that they were reluctant to report because, basically, they considered reporting to be a waste of time. There were two other groups, although smaller, who seemed to believe that the Company did not want or expect reports and that there was no interest from the Company. There are probably a number of very different reasons why certain seafarers felt so negative. Maybe it was a reflection of their own attitudes towards ISM or maybe it was a true reflection of a company who really didn't care. Clearly these groups are going to be the most difficult to persuade that reporting is a very important factor in the successful management of safety.

A reason for reluctance to report incidents which had not occurred to the author at the time of constructing the questionnaires was racial intimidation. It is not known how extensive this problem might be but the following was received from a Filipino rating working on board a Norwegian flagged vessel:

"There is a reluctance report anything because of racial discrimination — there is no democracy."

Certainly there were many comments received from OECD masters and officers which could be easily interpreted as having racist undertones. Reports from the Seafarers International Research Centre (SIRC) based at Cardiff University suggest that racism is not an issue at sea since the industry has achieved a relatively high level of globalisation. From reading some of the reports submitted, the author very much doubts that that reflects the actual situation.

Reluctance to report — perceptions of ship operators

Let us now consider how the ship operators responded to these issues. Remember that they are not expressing their own views about any reluctance they may have about reporting. Rather, they are giving their views on why they think masters and seafarers might be reluctant to report. In some respects they were quite accurate and mirrored the seafarers responses quite closely. In other respects they differed quite significantly

Certainly they recognised that seafarers did have concerns about possible repercussions with regard to their employment and career

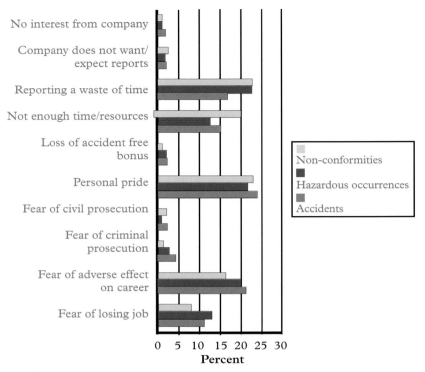

No interest from company

Company does not want/
expect reports

Reporting a waste of time

Not enough time/resources

Loss of accident free
bonus

Personal pride

Fear of civil prosecution

Fear of criminal
prosecution

Fear of adverse effect
on career

Fear of losing job

Non-conformities

Hazardous occurrences

Accidents

0   5   10   15   20   25   30
Percent

Figure 9.31 — Reluctance to report — ship operators

prospects. Somewhat surprisingly, the ship operators agreed that the masters and seafarers would not be unduly concerned about possible civil or criminal prosecutions.

One of the big differences came with personal pride. The ship operators perceived this to be a major factor that might inhibit reporting. The perception of nadequate time and other resources was also identified by the operators. The other major factor where the ship operators seemed to be even more pessimistic than the seafarers themselves was the idea that reporting was just a waste of time.

Interestingly, the ship operator respondents appear to have considered it almost inconceivable that the seafarer could think that the Company did not want or expect reports and certainly not entertain the idea that the Company might not be at all interested in any reporting procedure.

Reluctance to report — perceptions of other stakeholders

Finally, let us consider the perceptions of the other stakeholder group. What reluctance did they think might inhibit the master and seafarers from reporting? Interestingly, many of their perceptions coincided quite closely with the masters and seafarers own reservations. Perhaps the biggest difference was that the other stakeholder group put a much greater significance on the potential effect on employment and career prospects.

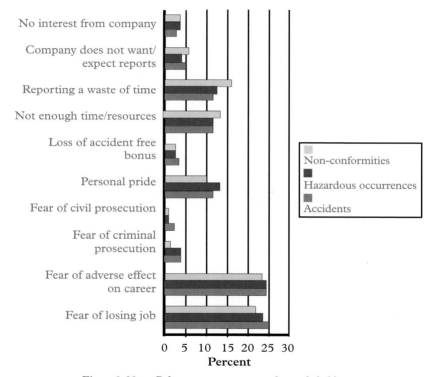

Figure 9.32 — Reluctance to report — other stakeholders

Generally the responses received from the masters and seafarers in this section seemed to be much more in line with many of the narrative comments that were received compared with the answers which had been provided earlier about the levels of reporting and ideas of no blame and safety cultures.

In an attempt to encourage seafarers to report incidents the Port State control officers are actively discouraged from close examination

189

of the actual contents of reports according to Captain Olle Wadmark of the Swedish Maritime Administration who said:

"the PSCO may ask for records of the most recent internal audits (but) should not normally scrutinise the content of any non-conformity notes"

The situation does appear to be that there is reluctance to report. To a large extent this reluctance is understandable, but if ISM is to work then we need to find ways of overcoming that reluctance. The issue was stated very well by a superintendent from a shipping company who said:

"The old concept of keeping everything wrapped up onboard and hidden from eyes outside needs to disappear and a system of transparency needs to be developed where crew are in a position where they can, with confidence and pride, report anything on board that is not right and needs to be rectified."

A related observation was submitted by an ISM consultant:

"Seafarers will only enter fully into the reporting process if they are confident of having the backing of management, including timely replies ad evidence of action taken. While many feel insecure in their jobs and are faced with either apathy or sometimes antagonism from management, one can understand their reluctance."

Without a doubt this whole issue of transparency within a no blame/ safety culture is likely to be difficult for some companies and individuals but, eventually, it will become 'the way we do things round here.'

Chapter 10

THE SURVEY PROBES OPINIONS —
IS ISM WORKING?

Introduction

At the very end of the questionnaire for masters, seafarers and ship operators and near the beginning of the version of the questionnaire for other stakeholders, two big questions were posed:

- In your view have the number of accidents, hazardous occurrences and non-conformities reduced since implementation of the ISM Code?

- In your view has the ISM Code achieved its objectives?

The way these questions were phrased was in anticipation of a somewhat subjective answer reflecting more a general perception by the individual respondent rather than any objective/factual statement. In hindsight the questions may, in certain circumstances, be flawed. This applies to the first question in particular.

The question includes a presupposition that there were occurrences of accidents, hazardous occurrences and non-conformities prior to ISM implementation that could be reduced. The second question makes an assumption that it is possible to make ships safer. A number of respondents seemed to hold the view that they were already operating close to perfection as far as safety was concerned and therefore the ISM Code had nothing to offer. The author can only apologise to those who were at that extreme end of perfection but, working on an assumption that there were probably not too many companies or ships operating without any accidents, not having a single hazardous occurrence and with such perfect systems in place that non-conformities just did not happen, then it is possible that the questions were not too seriously flawed.

The questions were intended to solicit a view from the individual respondents as to whether, in their experience, the ISM Code was starting to work/starting to manifest measurable results.

When the first wave of questionnaires started to be returned, mainly the blue master and seafarer forms, there was an unexpected and worrying surprise. The answers to these two big questions and the narrative comments that were included suggested a very negative attitude indeed to the whole concept of ISM. It had certainly been

191

anticipated that there would be some negativity but the apparent condemnation which was coming in was much worse than expected. Initially this took the author by surprise and it took some time before he started to understand what was happening.

As time went on, more and more positive and supportive replies were received and fewer and fewer outright negative responses. Still, the explanation for this strange phenomenon eluded the author. Before exploring that further though, let us consider the overall results of the survey.

Have incidents reduced since Phase One implementation?

The respondents were given the opportunity to choose from four specific statements. At one extreme they could say 'no definitely' or opt for the much softer, middle of the road 'no — not noticeably' and, at the other extreme end to report 'yes — significantly' to the weaker 'yes — slightly'. They could also declare that the ISM Code did not yet apply to their own situation and therefore they could not offer a point of view. They were also given the option to state that they did not know the answer. The analysis of the 3,000 questionnaires that had been submitted produced results as follows:

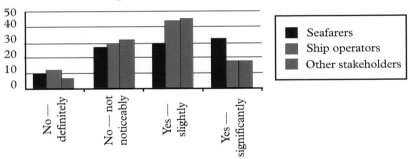

Figure 10.1 — Have incidents reduced since Phase One implementation?

On the whole there was a reasonable level of agreement between the three participating group, although the masters and seafarers were a little more biased towards the positive end. The overall result though was a fairly classic distribution curve with the bulk of the respondents somewhere near the middle, without strong views and a smaller number of strong views at each extreme end. The general view seemed to be that the number of incidents had probably not changed significantly since Phase One implementation, although more than 30% of masters and seafarers and a little under 20% of the other two categories considered that there had been significant reductions. In the end only 10% of respondents felt that there had definitely been no reduction of incidents.

Because of the way the database had been set up, it had been possible, without too much difficulty, to watch this curve develop during the period of the research. In the early stages the bias was very much towards the left, negative, side. Only in the later stages did the strong shift towards the right start to occur. It was not immediately apparent why this shift was occurring.

Has ISM achieved its objectives?

The next question, asking whether ISM had achieved its objectives, produced quite a similar result, although with a stronger bias towards the positive end of the scale. In hindsight, perhaps the question was too strongly worded. It made an assumption that it was possible for the ISM objectives to be achieved in the timescale since July 1998. The question, perhaps, should have been phrased in such a way that the respondent was being asked to comment on the trend towards achievement rather than having actually reached the goal. However, it would appear that few people took the question too literally and seem to have had the trend rather than the end result in mind.

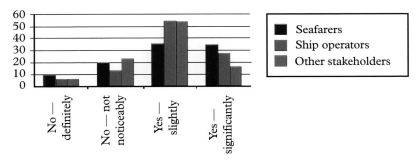

Figure 10.2 — Has ISM achieved its objectives?

Considering the answers to the two questions together it is interesting to see that the respondents seem to be suggesting that a reduction in incidents is not necessarily required in order to make progress towards achieving objectives. This is probably correct, in the early stages, since the setting up of the systems and, even more difficult, the setting up of a new way of working, i.e. no blame and safety culture, will not happen overnight. However, having considered results from companies who have started to approach the end of the tunnel, the rewards will start to follow as a natural consequence and accidents and claims will reduce.

Having said that, it is perhaps worth reflecting upon the views of some external observers. It is certainly possible that their views may be much more objective, since they should be impartial although they may not have the benefit of detailed knowledge of the full workings of the systems. A number of pilots submitted observations and, interestingly, many were quite critical of what they had seen. The following were two quite typical examples:

1   Working as a pilot (mainly tankers) I am not directly involved in the operation of the ISM Code. Boarding ships of various owners and nationalities gives a good opportunity to observe standards on board and get the views of masters and officers. Since the ISM Code was introduced there has been no noticeable improvement in standards. We see the same ships and the same people on them. All that has changed is that ship's staff are further burdened by a mass of paperwork. The success of the Code seems to depend on companies operating within the spirit of the Code but the companies which really needed the Code are hardly likely to enter into this spirit and just see the Code as another bureaucratic obstacle to overcome or circumvent.

2   As a marine pilot I see about 400 ships a year. ISM may be working on some ships but unfortunately, from a pilot's perspective, it seems many times to be a check-off ritual only. Often I see pilot cards as photocopies, with the bow thruster box ticked as fully operational when the ship does not even have one! Training still seems deficient in many ships — there is a perception that many personnel are just going through the motions. All the mate wants to do is obtain your name, sign the pilot card and whip it away from you, before you can even read it.

One pilot was based in north western Europe and the other in Australasia. The similarity of their observations and perceptions is therefore of even greater concern. It may be worth further research amongst pilots to establish exactly how widespread the pessimistic perception and poor experiences of ISM might be.

Cultural/national differences of perceptions

In the middle of 2001 the results of a BIMCO/ISF survey into manning was published (BIMCO/ISF 2000 Manpower Update — The worldwide demand for and supply of seafarers — Main Report — IER University of Warwick plus abridged report in ISF annual report 2001). The survey looked at the current situation with regard to where the masters, officers and crew who were manning the world fleet were presently coming from. It looked at the status of recruitment and training and made certain predictions about where the seafarers of the future would be coming from. One of the main purposes of the report, though, was to establish whether there was going to be an adequate and sufficient labour force in the future and, if not, what the shortfall was likely to be.

The author read the report as a matter of academic interest, not realising initially that it was going to have important relevance to his own research. The BIMCO/ISF survey split the world's seafarers into broad national/regional groups. Their findings suggested that the officers of the world fleet, split according to those groups, was as per the graph below:

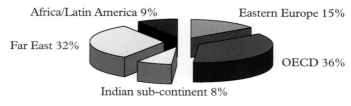

Figure 10.3 — BMCO/ISF study — officers

In the questionnaire the respondents had been asked to state their nationality and it occurred to the author to see if there might be any national/cultural differences to experiences and perception about ISM. It was somewhat labour intensive, slotting each individual into his or her correct group but the end result was well worth the effort.

The masters and seafarers who had responded to the survey and who had indicated their nationality, produced the following split:

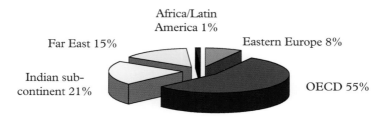

Figure 10.4 — ISM survey sample

This result did not map perfectly onto the BIMCO/ISF result. There was a greater number of OECD and Indian respondents and less Far East and eastern Europeans. However, it was felt to be reasonably close enough to the current world manning situation to be worth proceeding with an analysis.

Before considering the results of that analysis, it is also worth identifying another interesting and relevant factor which came out of the BIMCO/ISF survey. It had established that 50% of the masters and chief engineers, i.e. the command/most senior officer positions on board ships of the world fleet were still from the OECD countries, the other 50% coming from the developing nations — particularly India, the Philippines and the former Soviet Union countries.

OECD 50%   Other nationalities 50%

Figure 10.5 — Current position re masters and chief engineers

That fact in itself was not so important. What was much more significant was that almost all the OECD masters and chief engineers were over 50 years of age, who would be retiring within the next ten years and there were very few OECD nationals following behind to take their place when they did retire. From the late 1970s until the late 1990s very few potential officers were recruited or trained from OECD countries. The significance is that, within a decade, almost all the ships of the world will be commanded by Indians, Filipinos and individuals from the former Soviet Union. Of course there will also be masters and chief engineers from other developing nations.

In order to avoid too much clutter in the diagrams and to ensure a reasonable sample size, it was decided to compare the views of the three largest groups of respondents from the master and seafarer categories. These were the OECD nationals, the Indian subcontinent (mainly Indians) and those from the Far East, primarily Filipinos. When these groups were separated out the result was quite amazing:

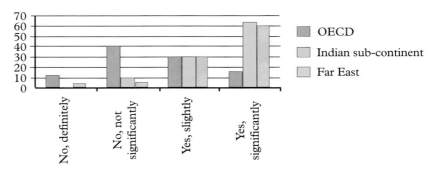

Figure 10.6 — Have incidents reduced since implementation? — a comparison of views

The responses from the OECD masters and seafarers shows a very classic distribution curve with a slight bias towards the negative, approximately 60% saying no; no — definitely or no — not significantly in answer to the question whether incidents had reduced since Phase One implementation. However, the results from the Indians and the Filipinos, which were almost identical, showed an enormous swing towards the positive side of the scale. They were not just going for the softer option of yes — slightly but over 60% had gone for the much stronger yes — significantly. Very few opted for the no — not significantly option and there were almost none who went for the very negative no — definitely end of the scale.

An almost identical picture emerged when the answers to the other big question were split in the same way — although the OECD respondents were shifting a little towards the positive.

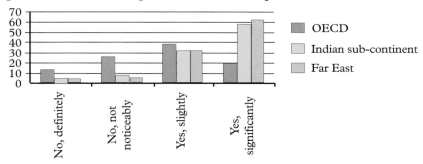

Figure 10.7 — Has ISM achieved its objectives? — comparison of views

The mystery of why the trend had been shifting during the course of the survey was starting to make sense. The initial distribution of the bulk of the questionnaires was with The Nautical Institute journal *SEAWAYS,* the *NUMAST Telegraph* and the *IFSMA bulletin.* As such, the first recipients would have been the shore-based seafarers in the UK. Whilst some of these may have been seafarers on leave or working on ferries or other short sea vessels, many would be retired seafarers or otherwise people who were working ashore but who felt that they did not fit into the ship operator or other stakeholder categories. The next wave of responses were from other OECD nationals, particularly from Australia, Canada, New Zealand and the United States who were, perhaps, from similar backgrounds. It was difficult to establish, in many cases, what their recent seagoing experience was or whether they had actually sailed with ISM systems. The group did however indicate a strong negative attitude towards the concept of the ISM

Code. It was some time later that much more varied responses started to be returned. Allowing for the time lapse, this was probably the deep sea seafarers who were actually working on board ships who were starting to get their questionnaires returned. As more time went by the completed questionnaires were increasingly coming in from non-OECD nationals, particularly Indian masters and officers and, later still, Filipinos.

If these results were accurate, i.e. that Indian and Filipino masters and seafarers were having a much more positive experience of ISM, or otherwise had a much more positive attitude towards it, then this would have, potentially, a very significant impact, bearing in mind the conclusions of the BIMCO/ISF survey. If the Indians and Filipinos were going to command and man a large section of the world fleet within the next few years and if they were so positive about ISM, this would surely provide much hope for the future successful development of the Code. The big question was; were the Indians and Filipinos being honest and accurate when they answered the questionnaire? This may sound like an unfair, rather arrogant and perhaps even racist question to raise. However, if the results of this survey were to stand up to the criticisms which were bound to follow, that question had to be asked and the possible answers fully explored.

There would appear to be at least four possible explanations:

1 The respondents were deliberately lying.
2 They had totally misunderstood the questions.
3 They were providing what they believed were the correct answers.
4 They really did believe that the ISM Code was working.

The author's views on these possibilities follow:

1. There were so many individuals involved that the idea of all those individuals each deciding to deliberately submit misleading answers has no logical justification at all. There is no evidence whatsoever that there was some sort of plot or conspiracy on the part of whole national groups to sabotage the survey by deliberately submitting misleading answers. There would be no purpose to be served by lying and any such idea is untenable.

2. Almost all Indian masters and officers speak and read excellent English and, invariably, have had the benefit of a good basic education before they embarked upon a seagoing

career. Whilst English is much more of a second language for Filipinos — their command of the language, particularly at senior levels, is very good. There is no justification therefore to suggest that the respondents did not understand the questions or the answers they were giving.

3. Whilst it is perhaps wrong to stereotype people — particularly by national groups — the author would venture to say that, in his experience, both Indians and Filipinos tend to be courteous and polite people. It does not seem to be within their character to want to deliberately cause offence — they would rather please. There are many Europeans who have sailed with Indian and Filipino seafarers who would tend to go even further in their psychological profiling of these groups. At its most basic they would say that Indians, more-so Filipinos, will tend to tell you what they think you want to hear. This is not done to deceive you but rather because the individual wants to please you or make you happy, whereas the truth would possibly make you worried or upset. Such an analysis may sound very patronising but it is a perception very widely held by Europeans. Could it be that the respondents, when they were working through the questionnaire, were looking for the correct answers, i.e. which answers would give most pleasure to the researcher and which answers would result in disappointment? The author has a number of highly respected and valued Indian friends. Some are still actively involved in seagoing careers while others are ex-seagoing and now working ashore in the shipping related industries. Their views were sought on this possible hypothesis, knowing that they would provide honest and objective answers. They all felt that there may be a little of that attitude of not wanting to cause offence by ticking the most appropriate box but, on the whole, they believed that the responses were accurate and could be relied upon. The author did not have similar direct contact with Filipinos but through intermediaries did solicit similar views and received similar answers.

4. This leaves the fourth and final option — that they might actually be telling the truth and actually believe that the ISM Code can work and is working. From reviewing

narrative comments received and discussing the issue further with many individuals it would appear that many Indian and Filipino masters and officers see the ISM Code as providing a structure — a framework — onto which they can build and secure their management systems. The way their systems are structured is such that the manuals tell them what they are supposed to do. They have access to and are directed towards relevant sources which tell them how to do it and they have a system that will tell them whether they have done the task or performed the process correctly.

These initial findings were published as part of a preliminary report. Interestingly, a European instructor who had been running a training centre in the Philippines for many years contacted the author with confirmation that he agreed with the basic conclusions. Even more interesting was an observation/suggestion he went on to make, that if a similar analysis was undertaken by splitting the groups by age rather than by national backgrounds then a very similar result would be achieved. The questionnaires did not ask the respondent to declare their age and therefore such an exercise was not possible. However, a review of some of the most negative responses received gave certain clues suggesting that the individual had been at sea for many years and in some cases had probably already retired.

The theory is supported by comments received from other respondents such as this Australian ship manager:

"My personal observation is that the old timers are slowly coming around whereas the younger seafarers are much more open and accepting of safety management principles, as it's covered at colleges and is also very much part of the working culture at least in other Australian industries"

If the theory of this European instructor from the Philippines is correct then again it creates additional hope for the future success of the ISM Code.

Another interesting experience was shared by a respondent who had first hand experience of both commercial shipping and the offshore industry. More than that though, within the context of this section, he had recent experience of providing safety-related training to non-OECD seafarers. He submitted the following report:

"I have just completed a training programme — where I essentially presented an ISM/SMS and FSA appreciation programme with a number of our Far Eastern and FSU affiliates and inspectors (predominantly Polish, Russian,

Ukrainian, Croatian and Filipino). Quite a number of these men/woman had recent experience as master and officer in a broad range of ships. There was almost unanimous support for the ISM/SMS regime throughout my courses. Their enthusiasm was striking. All gave evidence of its benefits and interestingly many noted that it facilitated greater interaction with owners/managers/superintendents. Again, without exception, they experienced worthwhile involvement of the management of their working environment.

Their enthusiasm unfortunately contrasts with the understanding of European seafarers as I have generally experienced it, the latter, on the whole, remaining sceptical of the benefits of ISM. I remember having similar feelings offshore before I had my 'road to Damascus", i.e. before I first recognised the real benefits of total safety management under the safety case regime offshore. Some of the cynics were shipmates when I was deck apprentice and officer with (various British shipping companies)."

As was suggested elsewhere — more research would appear to be warranted into the apparent differences in perceptions of ISM by various cultural groups.

Chapter 11

CONCLUSION AND THE WAY FORWARD

Introduction

Many of the findings of the survey could perhaps have been predicted in advance. Many companies and individuals are experiencing difficulties with their implementation of ISM. Paperwork and lack of resources are particular problems. However, the research has brought to light a number of new and relevant issues that can provide the basis on which future progress can perhaps be built. Some of the national/cultural issues are of particular potential interest. There were also areas of considerable concern which have been highlighted — serious misunderstandings about the Code which seem to be widely held by many seafarers — particularly from the more traditional maritime nations.

One fact which should never be lost sight of is that seafaring is, by its very nature, a dangerous occupation. An article appeared in the medical journal *The Lancet* on 17 August 2002, drawing attention to some recent research by Dr. Stephen Roberts at Oxford University. It concluded that fishermen and merchant seafarers have by far the most dangerous jobs in Britain. The researchers found that people working on the sea are up to 50 times more likely to die while working than any other occupation.

It is a sobering thought to realise that seafaring is not only more hazardous than construction and manufacturing industries but also a lot less safe than working for the police, army or fire brigade! Dr. Roberts analysed official death statistics from a range of different professions between 1976 and 1995. His top ten most dangerous jobs, in decreasing order of severity were:

1) Fishermen.
2) Merchant seafarers.
3) Aircraft flight deck officers.
4) Railway lengthmen (sic.).
5) Scaffolders.
6) Roofers and glaziers.
7) Forestry workers.
8) Quarry and other mine workers.
9) Dockers and stevedores.
10) Lorry drivers.

One thing that the survey confirmed is the very wide spectrum of compliance that exists across the industry. It would appear that most companies and ships which require DOCs and SMCs do have their pieces of paper but few would actually seem to have a functioning safety management system from which tangible benefits were being derived.

In a paper presented at the IMarEST Conference in May 2002 Brian Orrell of the UK based officer's union NUMAST made the following observation:

"We believe the implementation of the ISM Code today has merely served to confirm that the good ships are good, mediocre ships are mediocre and the bad ships remain bad."

A similar experience was reported by Jorg Langkabel of DNV who found that after two and a half years post Phase One implementation three types of companies emerged after certification:

- Those with perceived benefits (e.g. improved operational performance; reduced volumes of insurance claims, improved efficiency, reduced costs).
- Those simply meeting requirements in an average way.
- Those who struggle and/or could not see the purpose.

One further example of a very similar observation was received from a marine surveyor on the West coast of the USA:

"The better shipping companies have probably become better as the routines have been refined and as each occurrence has been investigated and action taken to prevent/minimize recurrence. The inferior companies have become worse, cutting corners with poorly trained crews and where the senior officers are counting the days to come ashore and/or retire. Commercial pressures to meet schedules in the North Pacific override good seamanship all too regularly. Container ships are scheduled to arrive at a terminal by the hour."

There are individuals who seem to hold the view that ISM is the greatest curse ever inflicted on the shipping industry whilst others believe that it has been the greatest blessing. There were many in between. What the survey has hopefully demonstrated is that questions such as whether or not the ISM Code is, or can, work are really not the appropriate questions to ask. The ISM Code is identical, word for word, for every ship operator, every ship and every seafarer around the world. What is different and what the questions should focus upon, is the individual, specific, safety management system. It is the SMS which differs from company to company and from ship to ship. It is quite clear from the results of the survey that there are some very good SMSs in place and working, thus producing some excellent results, apparently, including increased profitability. On the other hand, there would seem to be other SMSs which probably have no chance of ever working and which should be consigned to the ocean-deep.

A tanker master shared his nightmare experience:

"Questions answered basis last vessel 40,000 mt DWT product tanker. Absorbed into management five months ago. Previous operators — Greek. Now crewed by British captain and chief engineer, all others Ukrainian. None of the crew familiar with the Company. Vessel operated on an interim SMC provided within a leading classification society [name of society was provided]. Vessel in extremely poor condition, illegal in all aspects of MARPOL regulations due failed equipment. All class certificates valid despite serious defects. Vessel should not have been trading. Ukrainians do not have any concept of a safety ethos. Function of ISM was nil and impracticable. ISM detracts from safety on that vessel due to the time required of master."

A number of similar horror stories were sent to the author. Clearly a related matter of great concern also immediately arises as to why the seafarers, professional mariners, felt obliged to sail on such a ship. There was certainly no indication from the report that the situation was brought to the attention of the authorities. The suspicion would have to be that we are again back into the fear and blame culture in a big way.

It is interesting to consider the views of some external observers — such as this pilot who seems to have formed a most unfortunate view of the way in which many companies may have implemented their SMS:

"My overriding impression of ISM (and similar safe ship management systems for non-ISM ships) is that, with a few honourable exceptions, they are regarded as a means of demonstrating that shore based management has done their bit and that if anything goes wrong it must be the fault of the master and crew."

Much more needs to be done to track down the good SMSs — to analyse them and understand what it is about them that makes them good. How have those companies managed to motivate their team to embrace their SMS and fully implement it. When we read a claim from the chief executive of a shipping company such as this, we should hope that he and his company would be prepared to share some of their experiences and achievements for the benefit of the whole industry:

"Ships staff now understand that they are listened to and are eager to learn and do the best that they can. They own the ISM system and are part of the management team."

Perhaps we need to be able to quantify the improvements in efficiency and profitability to persuade some parts of the industry that it really is worthwhile making the commitment. There are many lessons that can be learnt from others within the shipping industry. Having said that, it is not the view of the author that any one company should slavishly copy the SMS of another company. The suggestion is that we can learn and understand the methodology adopted and see whether that can be applied to our own situation.

There would appear to be a need to improve the standard of education of some of the very concepts behind the ISM Code to many seafarers as well as ship operators — particularly to those who are perhaps outside of, or beyond, the formal learning systems, i.e. those who are unlikely to be attending college or other training establishments.

Further research will be necessary to understand fully the apparent major differences in perception of ISM between different national/ cultural groups. If the findings of this survey are confirmed then the opportunity must be taken to capture and develop the enthusiasm for ISM, as shown particularly by Indian and south east Asian seafarers.

Of course ISM cannot, and was never intended to be, a substitute for employing experienced and well qualified people. It was intended to complement such ideas in an industry which might have experienced a shortage of such individuals for a while. Whilst the author might not fully agree with everything this chief engineer says , he has certainly identified a key issue:

"ISM, I feel, has been introduced to bring standards up for flag of convenience vessels. For vessels already operating safely I feel it is another layer of inspections/ paperwork. Different standards between countries appear to exist. It is reducing the ability of the individual. Relying more on the system. At the end of the day it's the ability of the individual that counts and when people realise that to employ the right people will ultimately cost more but save money in the long run – then a safer environment will occur.

An investment in people is an investment in safety and an investment in both has to be extremely good for business: so what is the way forward . . .

The way forward

If I was to try and identify one overriding conclusion of the whole research it could possibly be summarised by saying that a very significant section of our industry still appears to be struggling to implement the ISM Code because of inadequately functioning safety management systems. Having said that it has also been established that there are examples of SMSs which can and do work. More than that when the SMS does work it results not only in safer ships but also more efficient ships and, of great importance, a more profitable operation altogether. That is language which should make any ship operator and his accountant, sit up and take notice.

The fact that almost every ship operator in the world has to comply with the requirements of the ISM Code is incontrovertible. They must obtain the DOCs for their office ashore and the SMCs for their ships. Inevitably resources must be made available both in terms of funding as well as labour to achieve this end. Such resources being poured in to an inefficient and ineffective SMS are, it is suggested, equivalent to pouring money down the drain. There may be pieces of paper, nicely framed, up on he wall and auditors may have been sufficiently satisfied to verify the paper system but it produces little benefit to the ship operator, the seafarers or anyone else. Such systems are potential major problems for ship operators — time bombs just waiting to be picked up by a PSC inspector who has acquired a proper grasp of what a working SMS should look like or a court or insurer who is interrogating the system in detail following an incident.

The fact is that the difference in levels of resources required to do what is sufficient to merely maintain a system that will just satisfy the external auditors to obtain DOCs and SMCs and those required to implement an SMS which is working well and returning profits for the Company is relatively small. Indeed it is suggested that in the short to medium term the Company which applies the additional resources and commitment will recover that extra outlay many times over.

The important point to understand is that the ISM Code is identical, word for word, for every ship operator and every ship in the world. The reason why there is such a diverse range of experiences of ISM implementation is directly attributable to the way in which the individual SMS was designed and put into practice. Some of the common factors identified with SMSs which did not appear to be working satisfactorily, or at least where considerable negativity was expressed by individuals involved in its implementation, are set out below:

*   Too much paperwork.
*   Voluminous procedures manuals.
*   Irrelevant procedures.
*   Bought-off-the-shelf systems.
*   No feeling of involvement in the system.
*   Ticking boxes in checklists (without carrying out the required task).

* Not enough people to undertake all the extra work involved.
* Not enough time to undertake all the extra work involved.
* Inadequately trained people.
* Inadequately motivated people.
* No support from the Company.
* No perceived benefit compared with the input required.
* ISM is just a paperwork exercise.
* No respect for external auditors.
* No respect for classification societies.
* No respect for Port State control inspectors.
* No respect for the shore management by the seafarers.
* No respect for the seafarers by the shore management.

What must be understood is that it is not the fault of those people who expressed these negative attitudes that they feel the way that they do. They are the unfortunate recipients of a concept which has been basically dumped on them with little or no preparation, training or involvement. It is little wonder that so many systems produce little, if anything, of any real value. The SMS can only work if those who are involved in its implementation actually want it to work. This is at the heart of the very nature of management systems and is what differentiates them from prescriptive rules and regulations.

The motivation required to link the individual to the system can only arise within the concept of a culture. At its advanced stage it manifests itself as a safety culture but at the earlier stages it appears as a company culture. These are not just flowery words of modern management speak — it is the reality which must be understood if progress is to be made with ISM implementation. To highlight this point it is perhaps worth comparing some of the common factors which appear in those companies who appear to be operating very successful SMSs with the above rather negative list.

When an analysis was undertaken on those companies who claimed to be experiencing very positive results from their ISM implementation there were indeed a number of factors which were common to all and which stood out as being of special relevance. Some of these are set out below:

* Leadership and commitment from the very top of the organisation, i.e. from the shipowner, chief executive, managing

director, etc. and from that commitment and leadership throughout the management structure.

* Paperwork reduced to manageable levels — including procedures manuals, checklists, reports, etc.

* A sense of ownership/empowerment by those actually involved in the implementation process of the SMS — i.e. the personnel on board the ship.

* Continuity of employment of personnel both ashore and on board ship.

* Two way communication between ship and office — with mutual respect.

* Awareness of the importance to the individual and to the Company of managing safety.

It was out these various attributes that the Company culture, and in turn the safety culture, flowed as a natural consequence. The whole process of managing safety is not something which stands in some sort of isolation it is just part of the way in which things are done in that particular company. When these various components are combined they seem to be sufficient to produce a working environment in which people take responsibility for their own safety and contribute towards the safety of others and the Company as a whole. As a natural consequence of that shift in attitudes and values accidents, incidents and, consequently, insurance claims all start reducing. Once that starts happening there is a much more efficient use of time which allows genuine efficiencies to be made with the consequence that less money is draining out of the Company.

Whilst I could see this, from my capacity as a researcher, I questioned my ability to be able to adequately and convincingly communicate the full significance of the message to third parties — particularly the hundreds of individuals who had reported such terrible personal experiences of ISM implementation. I have not personally sailed with a SMS which had been set up pursuant to the requirements of the ISM Code. I have not been directly involved in constructing and running an SMS. I have had the opportunity of looking at numerous individual safety management systems, of considering hundreds of individual experiences and analysing factors which seem to affect how an SMS actually functions in practice. However, that does not, in my view, qualify me to argue a strong case for what needs to be done to actually implement a good, working and efficient SMS.

Such a position could only be viably argued from individuals with first hand, direct experience. Individuals who could stand up as living testimony that it can be done.

Accordingly I looked at the list of organisations and individuals who had been in contact with me and who claimed to have had positive experiences of ISM implementation and to try and persuade a diverse selection to share their experiences in some detail. For practical reasons the number of such individuals had to be kept relatively small — I therefore identified one designated person, representing the shore-based part of the implementation process, one ship master able to consider the onboard issues and a training provider who was able to provide a more general overview. The three individuals agreed to share their experiences by writing an additional chapter each to form an integral part of this book. In this way they would provide some possible answers, from first hand experience, to some of the problems I had identified.

The three were complete strangers to each other but it quickly became apparent that they were all singing from an identical hymn sheet. Through their individual experiences they had each quite independently reached very similar conclusions and solutions to the requirements of what is needed to develop a successful safety management system.

The final part of this book therefore comprises those three additional chapters. It is not intended that anyone should try and slavishly copy exactly the experiences of these individuals. It is a very important feature of ISM that each company must develop its own SMS to reflect the way they manage safety in their company. However, there are general principles which should be understood and lessons which can be learnt. Armed with those general principles it should be possible to understand what needs to be done to change the attitudes, beliefs and values such that there is a sufficient culture change to see the SMS start working efficiently, effectively and profitably.

It would be appropriate to say a few words about each of the individuals before handing over to them:

Captain Stuart Nicholls had been in command at sea with a first class liner operator, where he had helped introduce their SMS and believed that (like many seafarers) they managed safety perfectly well before ISM and that it really added little benefit to their operation. However, he then underwent a career change and went to work onboard an offshore mobile drilling unit. He there encountered a level of safety management far beyond anything he had ever imagined

— where safety really was raised to the very highest priority beyond everything else. He does not advocate that the way safety is managed in the offshore industry should be imposed lock-stock-and-barrel on the merchant marine but he does suggest that there are many lessons we can learn from that industry.

Captain John Wright has vast experience of shipping as well as offshore and other industries and provides management skills training to senior officers in the merchant marine, in particular Crew Resource Management (CRM) as well as risk management. He has almost daily encounters with a wide range of seagoing masters and officers with whom he explores management techniques which might currently exist on board many ships and he teaches the type of management technique which is anticipated by management systems such as the ISM Code.

Captain Sean Noonan was a DPA in a relatively small shipping company who had set up one of the most imaginative and successful SMS I had encountered throughout my research. As a direct consequence of their ISM implementation the Company saved enormous amounts of money. They devised an SMS which plugged the many holes which existed in their organisation through which losses were pouring out.

Since each of these three individuals had their own fascinating story to tell, from first hand direct experience, I felt that they should be given the opportunity to tell their own story in their own words rather than for me to provide a second hand account. I am most grateful to each of them that they very kindly agreed to contribute their own story by way of additional chapters to follow my own findings and conclusions in this book.

Chapter 12

SAFETY MANAGEMENT IN THE OFFSHORE INDUSTRY — A QUESTION OF BELIEF

by *Captain Stuart Nicholls MNI*

Foreword

This is a message from an enlightened employer to his people:

"Gentlemen, if you forget everything else that has been said remember this. The most important contract you will make is not with yourself, your colleagues, your employer or even your bank manager. It is with your family — your mother, father, husband, wife, sons or daughters whoever it may be. You have committed to them when you say "I will see you soon." My responsibility is to make sure you can carry out that contract every time you go to the work place. Thank you gentlemen for helping us help you"

Why do we need this ISM system?

Systems are used by people to manage business, the process of making a profit. Our business is to transport goods safely and efficiently from port to port. The more we transport the more profit we make. However, transport brings with it many costs. Costs are deductibles, more deductions brings less profit. Businesses which have good systems are likely to produce healthy profits. Businesses with bad systems fail to maximise their potential profit. The ISM Code is a good system and has an important role to play in this climate of change within the shipping world. It is no coincidence that companies who manage health and safety well are companies who successfully manage their entire enterprise well and profitably.

With all operators required to comply with the ISM Code, everybody should be making good profits, given a reasonable economic climate. So why is that not the case? The simple answer is, systems are used by people. People are humans and where there are humans there are mistakes and omissions. The fashionable term is human factors or, more to the point, human failures. We need to reduce this human element to close the loop so as to benefit from a system that will reward, not inconvenience the ship operator.

Other chapters in this book explain the ISM Code in detail, from concept through implementation to the present state. The research and management experience brought to bear in those chapters contrast the benefits of the Code with the way seafarers and operators have received the initial ISM years of implementation with scepticism.

This chapter shows how the offshore oil and gas industry has successfully reduced the human factor since the *Piper Alpha* oil platform disaster in the North Sea in July 1988. It is an industry that has gone a long way to answering the same problems faced by today's shipping industry. It quickly identified the importance of closing the human factor loop and placing priority on people who use the safety systems. To achieve this, staff were trained in non-technical skills to help control risk of injury to people, harm to the environment, damage to property and loss of productivity. These non-technical skills are well received and practised offshore.

My illustrations in this chapter are based on my experience working within the oil and gas industry comparing it with my varied seafaring career which culminated in command experience

By describing how the offshore industry has profited from these systems, by focusing energy and resources on training the front line manpower, I believe, endorses the ISM Code for what it is, a well founded guide pointing the shipping industry in the right direction. I believe it is now up to the senior management of individual ship operators to pick up the baton and place their energy and resources into sea staff training. Then, they too will enjoy the same successes in the future as our offshore cousins.

The story of contrasts

My career started with a reputable shipping line with a large fleet trading worldwide. The Company on the whole was a thoroughly professional organisation, well managed with loyal sea staff, reflecting commitment and honesty from management. I followed the traditional method of working with the deck crew during the formative years gradually rising to occasional visits to the bridge. This was complemented with studying the theory behind the practice at college. After completing the initial training period I was introduced to the officer of the watch position under supervision and then, upon joining my first vessel as an extra deck officer (4th officer), I was guided through difficult passages allowing confidence to grow.

The fleet was made up of a single nationality. Crew levels were high with radio officers, chief stewards and leading stewards and on frequent occasions as mentioned above, a 4th officer. The master was a man rarely seen, rarely heard and considered by his crew to be somebody who sat at the right hand of God himself. The vessel was run in a very autocratic manner, with communication disseminated down through the ranks from the master and filtered up the ranks, leaving the chief officer to inform the master about daily matters. There was little one to one communication with the master at a junior rank.

My first encounter with the ISM Code was during my time as third mate, when I was asked to write down in my own words how I carried out an anchor operation. The master requesting this information could only give a brief explanation, saying it was some new fangled idea made up by shore management to justify their existence. The second mate was also requested to complete his version of chart correcting procedures. So we went about writing basic procedures for all our responsibilities at our level.

Unknown to us, every vessel and every rank was asked to record their methods. These were filtered and best practices from all was implemented into a company 'Code of Practice Manual.' Looking back, this approach from shore management to assess how the front line operators carry out their work was very reasonable, practicable and professional.

Today they call it empowerment or ownership of the system. Unfortunately we did not sense the empowerment because the master was unaware of the reasoning and this filtered down to the juniors as a waste of time. How far from the truth that was. An excellent system fell at the final hurdle through on board management style and poor communication throughout.

From the start the Code had a low profile entry into the industry, which was reinforced during my time at college studying for my chief officer's certificate of competency. During this time I read an advertisement in *Lloyds List* asking for people with relative ISM Code experience and, unsure of the prerequisites for the job, I requested help from a senior lecturer who admitted he had never heard of the ISM Code. That was November 1994.

After a successful time under the British Red Ensign I wanted to experience various ships and trades and sailing on flag of convenience vessels became a viable option. After a very limited interview that comprised questions concerning my knowledge of paint schemes and whether I would be available to join immediately, I joined my first ship operator under the FOC as chief officer. The Company was an internationally known food producer with vessels as part of its global logistics. Ships were seen as a necessary hindrance that cost money and were considered badly run in comparison to other modes of transport within the corporation. My first appointment with my new employer was a vessel in dry-dock. I passed the outgoing chief officer on the gangway and after uttering his disapproval of the operators inability to organise he jumped into the taxi I had arrived in and was away.

Unhappy with my familiarisation to the Company and its procedures I attempted to extract information from the master. He, along with most of the other staff made up of many nationalities, was not forthcoming . How was I going to survive the initial learning period? I remembered the Code of Best Practices and, unaware of phase dates for ISM Code implementation, just assumed the vessel would have one. Luckily it did. I then had the pleasure of taking the Code out of its packaging, read the pertinent chapters and as a result gained some understanding of my responsibilities. This is how I managed to survive the onslaught of requests for paperwork from the office.

Shortly after joining the Company we had our first internal audit for ISM compliance. This was led by a master within the Company and he was assisted by the director of procurement. This particular vessel was the first within the fleet to apply for a Document of Compliance. During the opening meeting there was a very bewildered look on everyone's face, which silently said "*What in God's name is the ISM Code?*" As a result, the opening meeting was used by the auditor to explain what the ISM Code was about.

The master saw the audit as an invasion of his power and began to condition the shipboard staff into thinking the audit was a waste of time. He began to ridicule the auditor and his assistant and went to great lengths to explain to all shipboard staff that the only way was his way and that we were to forget about what these office wallahs were preaching. He said they were not that bothered about procedures for this and procedures for that in dry dock, so why should we be bothered now. These words would demonstrate their significance to me later when I entered the oil industry.

The audit was short and to the point in my area of responsibility. Observations and non–conformities were raised, not surprisingly, given my short period within the Company. I used the observations for what they were, discovery of errors within the system, not reflections on my competence. The master, however, did not see it that way. His voice was getting louder and louder as the audit went on. He said "*What gives you the God given right to come in here and tell me how to do my job. I thought you would have been better, but no, now you are in the office you are just like the rest of them and think you know everything!"* The audit seemed to go on and on. However, over the course of the day the master slowly came around to what the audit process was actually trying to achieve and, sensing it was not him they were auditing, requested where he could find all this information. The auditor pointed to the pile of unopened manuals on the office desk.

My opinion on the flawed manner in which the ISM Code had been introduced was finally sealed when I was completing my studies for master's orals. It was the year of Phase One implementation and ISM was the buzzword around the questions and answers circuit for students. Buzzword was the right description because there was very little in the way of available information in the college nor were there any lecturers who could define or understand the spirit of the Code sufficiently to teach the students. Needless to say, under Murphy's Law, during my oral examination, when nearing the end, the examiner asked me to explain in my own words the ISM Code and, in particular, risk assessment.

To do this I used my recent shipboard experiences to explain the Code, quoting controlled documents, the master's responsibilities, personal injury, damage to the environment, etc. Happy with my answer he pressed for the risk assessment part and fudging together college notes and Code of Safe Working Practice notes I gave a very vague answer. The examiner was expressionless and turned to the window and took my notes. While I waited I really thought I had blown the entire exam. Still looking at me he gave nothing away until he wrote across my certificate the word 'passed' and then said, "You know about as much as I do on all this risk nonsense". From that moment I knew the industry was not going to do the ISM Code justice.

Taking command should be the proudest moment of your career. It was, but it was soon soured. I travelled all day to join the vessel in Europe, arriving late evening. The outgoing master expressed his concern for a speedy handover so that he could complete the procedure prior to an ISM external audit the following day. We completed late that night and I was up early again to prepare for the audit.

The audit was not reasonable. The class auditor had a non seafaring background and questioned everything in the manual. He was looking for documentary evidence on impractical issues and there was heated debate on certain non-conformities raised. The heat came not just from the auditor but also from the shore staff present during the audit. With the outgoing master paid off and away I felt that apportioning blame to myself and the outgoing master was undeserved and unnecessary. The shore staff saw the non-conformities as a bad result against the vessel, which directly reflected on the ship's superintendent and shore technical staff.

My position as newly promoted master was soon tested. The weather deteriorated during the audit and fog closed in. There was a long river passage to follow a difficult harbour manoeuvre. I insisted that the short handover, audit and weather situation justified a rescheduled sailing time but my decision was overruled. There was no written instruction to sail, only a discreet phone call to remind me of my new position. The message was simple "Keep low and no one will hear about you and the job is yours to keep." The seed was planted for an early departure from an industry I loved.

"Training? Training costs money!" That old gem was wheeled out on every occasion I requested a training course for myself or for the ship's staff. For example, I requested that the chief officer should complete some type of approved safety officer training. My employer was quick to reply that this was the responsibility of his manning agency and a matter for him to take up when he got home. If that wasn't enough, I thought I could attend the course myself and then train the crew on board but I was informed that there was no money and all information necessary to complete this duty was contained within the Code of Safe Working Practice Manual.

When I had a number of years experience as master under my belt, my principles were pushed to the limit. On arrival in a large US port with minimum turn round time, my senior director of ship management came on board to conduct his annual inspection. He was closely followed by the class surveyor who was to conduct our annual safety and construction inspection. To stretch our minimum manned vessel to the limit, the customs then arrived for a full scale inspection and, when I thought it could not get worse, the US Coast Guard, wearing their Port State Control inspector's hat, boarded and demanded my immediate attention. Not wishing to upset any of these people, with my senior staff we somehow managed to satisfy everybody in a way only seamen can.

The final straw came when we were ready to sail and the class inspector noticed a condition of class on the lifeboats. Pressure was mounting; the technical staff at head office were adamant the ship would sail with this condition but the class surveyor was equally adamant we would not, unless we corrected the condition prior to sailing. The charterer, who was standing in front of me, was also adamant we would sail immediately. Nobody was moving and the master was in the middle. I was instructed to use all means to persuade the surveyor to credit the condition, while my technical manager would contact the class head office and convince the surveyor's superiors that this was not what my company expected from a loyal customer. Phone calls brought about intense pressure on everybody and finally it was the class society that broke. I convinced the surveyor we had a long river passage and that we could take a repair specialist with us and land him with the pilot. Provided I sent pictures of the repairs via email the class surveyor agreed we would have no condition. We sailed on time with no conditions and the last call I received from the superintendent in the office was cold and calculated.

He said "Well done captain. It's not how you get into trouble, but how you can get the ship out of it. You certainly showed you could get out of it."

I could record here many such incidents of a similar description, which are not uncommon in today's world of shipping. The above anecdote only cements my belief that the ISM Code will not survive, never mind grow, if there is no understanding of the requirements and responsibilities by shore staff. When I say responsibilities, I mean standing by them, not paying lip service to them when an audit requires them too. This will not happen until the business benefits of managing safety are recognised and the industry culture changes to embrace fully the spirit of the Code's intent.

With this attitude in mind and the mind of other ship operators I encountered in my career I left the shipping world. I left when everything was going well for me. Master, office secondment imminent, so why throw it away? At the end of the day I did not believe in what I was doing and did not believe in what management teams ashore were preaching. I knew that if they did not believe in the importance of effective safety management and the effective implementation of the ISM Code, that it would catch them out one day. I knew that people could die as a result and they were welcome to my share of that!

Shipping will always comply with legislative requirements. That is the culture of the shipping world. The ISM Code is a requirement and all operators by July 2002 had to demonstrate they had the systems in place. It is clear from the audit process that people are operating the system but unfortunately observations show we are not operating the system to its true potential. Worse still, I do not believe the work force of the shipping industry has bought into the true spirit of the system, which is to reduce incidents; incidents that attract unnecessary costs and media attention to the shipping world.

The contrasting story

Here is a quote from a manager in one of the world's largest offshore drilling contractors when he was speaking to his people:

"What is our number one priority as one of the world's largest drilling companies?" Silence throughout the room. *"Anybody?"* Silence again. *"Safety "*. *"Wrong, safety only costs money"*. *"Environment"*. *"Wrong again"*. *"Employees"*. *"Wrong"*. *"Money!"*. *"Almost right"*. *"Profit"*. *"Damn close, keep going"*. *"Shareholders"*. *"Bingo. Is everybody happy with that, is everybody happy that the shareholder is our number one priority"*. The silence was starting to show on the man's face. A voice came from the back of the room. *"No, I am not. Why am I not your priority and my safety while I work for you"*. *"Thank you, there is somebody who is unhappy. Gentlemen we cannot make money for the shareholders if you are injured or damage property or the environment. So we have to make all the above a priority. When safety and money have a conflict of interests, safety of personnel, environment and equipment will take priority in that order. Is everybody happy now?"* Heads were shaking with approval.

What a way to be introduced to an employer and a new industry on your first day! It felt like I was being sold the Company and it felt like a breath of fresh air. This was an operator that talked the talk. My reservation of whether they could walk the walk would be answered later. Right now I was enjoying some serious employee/employer bonding.

The bonding was immediate. 0800 on a Monday morning and we were being introduced on first name terms to a whole host of senior management, not least to the vice president himself. He sat down took a coffee and chatted for a few minutes about where we, the people on the induction course, were previously employed . He focused on the skills we were bringing to him and how we would all play a vital role in the Company. To this day it was possibly the best bit of man management I have ever witnessed. The week was broken up by these management visits and they certainly gave me a very positive feel about my new employer.

I, along with many other new recruits, enjoyed the week's induction process, which was military in its precision. All documentation was completed with help at hand, all certification checked and double checked. We were even supplied with a list of things we needed to take out to the rig and for people new to working away this was invaluable advise and a reflection of careful thought being applied to the induction process. Other helpful inclusions were; pension decisions were made easier by experts brought in to show the benefits of differing schemes; health insurance benefits were described and expense forms for travel and training days were explained.

During the induction a training matrix was laid out in front of me and I was asked to look at my job description and associated training programme. I looked at the amount of training and the subjects covered and I was stunned. Subjects included management, safety, leadership, teamwork and communication skills. The list went on and on. I was bowled over when I was informed that if I wished I could approach my rig manager and request other courses not listed in the matrix under my job description. I was further surprised when told that I could apply for training of any type, provided benefit accrued to both employee and employer from the training.

After completing the formalities of contracts and job specific responsibilities, all the new recruits were introduced to the safety ethics of the Company. This was not a simple declaration of a mission statement followed by the health and safety policy being forced down our throats. Instead, the Company required new recruits to talk about their beliefs in safety and how important they perceived safety to be in the work place. As previously stated, the induction process was well thought out and the Company clearly did not want to lose the momentum with this topic.

Small debates started within the group and the group split into those who believed that safety was working in their previous employment and those who believed in theory but were negative as to the results they saw during their previous experience. I sat there still wondering which camp I should side with. My heart said it works, my head said it does not. I looked around and saw that the group of 20 people who had bonded well during the first couple of days were now on the point of rebellion due to their differences of opinion regarding safety.

Finally, the course coordinator interjected. We spent the next two days ironing out our preconceived ideas and support was given throughout this process by members of the senior management team who appeared regularly. Slowly, one by one, we were coming round to believing in the company's beliefs about safety management.

During our final day we concentrated on the practicalities of risk assessment and permit to work processes. The coordinator set small scenarios to help us to understand the relevance of the Company safety systems and how to administer them. Recruits who had no experience of these systems were given extra tuition and confidence grew within the group as the process came to a close.

Before we left we were introduced to the Designated Person Ashore. He began his speech with references to his experience working in the field and how, in the beginning, he had resented the inconvenience of safety systems because they meant nothing to him. He recited the old culture of rush, rush, rush indicated by the words *"No time for that crap, move on, time is money, go, go, go, go. Forget this, forget that. If you don't like it move on somebody will fill those shoes, boy"*

As he progressed with his absorbing anecdotes I began to think back to what I had left behind in the marine industry. The DPA's speech gathering pace as he started described how he had seen himself slowly turning into a dinosaur. He said *"People were not listening. They wanted change and Piper Alpha made sure we all changed. No more rush rush. We needed to take time out for safety, to plan the job, to use permits to work, to carry out risk assessment. These things all started to arrive but still meant nothing. Why? Because we did not have a culture of safety. We were still the same guys out there, except that now we were filling in paperwork to say we were safe. I will say it clearly now; the paperwork is useless unless the person filling it in believes in the system."*

He progressed further into how the Company had invested time and money into changing peoples work cultures. This is when he delivered his closing thought, with which I began this chapter, as to why we needed safety culture:

"Gentlemen, if you forget everything else that has been said this week, remember this. The most important contract you will make is not your commitment to safety, your colleagues, your employer or even your bank manager. It is to your family, your mother, father, husband, wife, sons or daughters, whoever it may be. You have committed to them when you say "I will see you soon. My responsibility is to make sure you can carry out that contract every time you go to the work place. If at any time you believe you will breach that commitment because of work practices being carried out on board, your unit can then call me! Thank you, gentlemen, for helping us to help you."

Now I was to experience whether the Company could walk the walk regarding the safety message as well as they could talk the talk. Arriving on board my first drill rig we were advised of a familiarisation meeting to be held in half an hour. The meeting introduced the offshore installation manager (master) and company man (charterer). We were briefed on the present state of operations and future developments and the OIM completed his introduction by reiterating his belief in the safety culture and the tools used to maintain the safety systems.

We were then shown around the rig safety equipment and our boat stations and our individual responsibilities were explained. The familiarisation process did not stop there. All personnel new to the industry were given a red hard hat and personnel new to the rig (but not to the industry) were given green hard hats. Core experienced personnel collected their white hard hats. I felt very uncomfortable being made conspicuous by my inexperience and what some might see as my incompetence. However, my mind was put at ease when we were reminded that it was to identify personnel who had not completed the safety training and to allow others to look out for us.

The rig tour culminated in a demonstration of the tools used on board to maintain the safety system. They were "at risk" behaviour cards, mini risk assessments, job specific risk assessments and permits to work. I was familiar with these tools following the induction period and served to reinforce the message that they were indeed used.

"Excuse me mate, can I have a word?" Embarrassed and nervous I went over to the man requesting my attention. *"Yes sure, how can I help you?"* I replied. *"Don't worry, there's no problem, I was just walking past and saw you doing a good job, but noticed you weren't wearing your safety glasses"* said the man in a confident manner. *"Yeah, I couldn't see what I was doing, so I took them off"*. *"Sure, no problem, I understand, but what do you think could be the worse thing that could happen doing that job?"*

Wait a minute I thought, who the hell is this guy, asking all these stupid questions. For goodness sake, I know what is safe and what is not. I certainly resent somebody whose name I do not even know, telling me what to do, or worse making out I was unsafe. Under duress I answered the man, *"I suppose I could get flying debris in my eye."* *"So best keep them on, after all you want to go home with full sight, yeah?"* *"Yeah, cheers mate."* *"No problem, we have an agreement then, you will keep them on?"* *"Yeah"*. *"OK then, mate, see you later."*

That was a very uncomfortable experience on my first day. I had never really experienced such a direct approach from a colleague before. Little did I know that the man then recorded his conversation on a card which was placed with many more that day. I had heard of this system before on ships and it was blessed with the nickname of 'shop a shipmate'. I can now see why it did not survive back on those ships, because it was introduced unsupported by the necessary safety culture.

Over my first trip I utilised the Permit To Work (PTW) system and the risk assessment systems with some ease but saw very little benefit from a safety point of view. I simply filled in the forms and carried on with my job within the marine department. It was the At Risk Behaviour Cards (ARBC) that made me feel uncomfortable since I was not used to the system and certainly not the theory behind it. I voiced my concern to the OIM who assured me that I would gain a full understanding after I had received the training in safety culture next time on leave.

"*Who believes in safety?*" Here we go again, I thought, more bloody safety. The silence returned from my induction process. "*I see no believers, that's normal on the first day.*" The course tutor described how we would all come to believe and use the safety systems within the Company. Looking around at the audience I was not convinced. Still, I was willing to be proved wrong. "*OK gents, in front of you are two pieces of paper. On the first is a picture of a tombstone, the second is a telephone. That is your tombstone, you have been killed at work, please write what you would like on your tombstone. The phone is to represent the phone call you have to make to family of your colleague who has just been killed at work. Please write what you are going to say.*"

I was a little stunned. Still I managed to use some of my experience as master to put together a reasonable telephone conversation. The tombstone was not easy. I had unfinished business in this life and I was not happy I was writing this obituary so soon. The message was simple and clear. People die at work when they should not. If we thought we felt uncomfortable after completing this task, it was to get worse.

"*OK, none of you want to die, but people do die in this industry. So how many is a fair number within the entire drilling industry, do you think?*" I could not believe it when I heard numbers between one and 10. The message was hitting home with everybody. Nobody wanted to die but how do we stop this? Understanding how fatalities occur would help us to reduce the chances of an accident happening and enable us to achieve the ultimate goal of an injury free work place.

The incident triangle, shown elsewhere in this book, shows how every fatality has a build up of lost time incidents, medical first aid cases, near misses and finally at risk behaviour. Simple when you think about it — look at the bottom and stop it before it gets to the top. The course continued to develop the skills necessary to allow everybody to conduct an at risk behaviour conversation and, most importantly, to get an agreement with the person with whom you have spoken. This, of course, had been what I had experienced on my first day on board and I realised that the card that had been completed afterwards was simply recording the at risk behaviour and was not a criticism of me as an individual.

Commitment from management is important in any system implementation and safety is no exception. We had the equivalent to ship's superintendents on the course from start to finish and the vice president made an appearance on the last day to reinforce the course material. He was emphatic when he stated no man would be held accountable if he stopped the job for safety reasons. Even the most sceptical of personnel on the course, some of whom had been drilling for 30 years, had been convinced the Company was serious about safety.

Back on board the rig I built up my confidence in reporting conversations I was having with people regarding their safety. Soon it became apparent I was helping with the safety culture. I now clearly understood the safety systems on board like job risk assessments, mini risk assessments and permits to work. When simple steps were insufficient, time was taken to reassess the job to make it even safer.

The weekly safety meeting is held by the OIM in front of everybody on board with no exceptions. He addresses alerts that have been received from other company rigs and rigs outside our own company. I realised that all companies in the industry had the same approach to safety and that there was a genuine safety culture where everybody in the industry was willing to alert each other to prevent further incidents. After the OIM has delivered his message the floor is opened for comments from everybody and this facility is well utilised. I was amazed at the level of some people's safety and environmental knowledge. These meetings are very worthwhile forums.

The meeting closed on a light note by announcing the at risk behaviour card of the week. The winner's cards has to show all the prerequisites of the training given, with the at risk behaviour being identified through the conversation with the person involved and an agreement reached to stop and change the behaviour and/or rearrange the work place. The OIM and the safety facilitator choose the card and the winner is announced after everybody has been informed of the trends from the cards received that week. The winner receives a quality gift, but importantly not money, as this sends out the wrong signals.

Trends come from the cards, since they highlight common problems. This can help management to campaign in certain areas of safety behaviour and also provides a foundation for near miss reporting. As per the accident triangle, one can generally see a trend growing towards a high potential near miss and this presents the chance to take action before the near miss event occurs. The facilitator's task is to follow up on the trends and readdress safety practices within the work force to prevent a near miss, life threatening incident or fatality occurring.

As stated before, the Company man (charterer's representative) remains on board at all times. He notes the amount of cards received each day and compiles a report to head office. Here is where due care is necessary so as not to fall foul of your own success. The rig had generated 13,000 cards in a 20 month period and with an average crew of 80 persons that equates to two cards per person per week. This was an excellent reflection of a good, honest reporting system and huge improvements were being made on a continuous basis. However, the report compiled by the Company man showed our rig to have the worst record in the entire industry, worldwide. How could this be? The consequences were astonishing. The charterer expected the highest standards of safety on all its chartered vessels and we were bottom of their list and our charter looked to be over. The crew were becoming restless and the reporting dramatically fell away as they believed they had been betrayed by the system they had bought into. Finally, sense prevailed, when it was noted that the computer system used by the charterer for analysing the data from the reporting cards was giving inaccurate responses to vessels which were reporting honestly and openly. Computations can be made from cards but human intervention is the simplest and most effective. We knew we could only learn and improve if we gathered in the learning opportunities presented from our data. We further knew that these opportunities would disappear if the trust we had built up was jeopardised and the reporting stopped.

There comes a time with all systems when you either believe in it or you just go along with it. I really bought into the reporting system when the vessel entered dry dock in Europe. The reader is invited to contrast this experience with the one described previously when I was on a conventional ship in dry dock.

The dry dock had a reputation for low standards of safety and our company understood the implications of the dock's substandard systems for our rig and its personnel. As part of the commercial agreement the dock management team agreed to put all their workers through an induction process into our company safety systems but when the rig arrived a week later it was evident that induction training had only been conducted for supervisors and above.

The dock employed a lot of contract personnel, with many being immigrants from Eastern Europe. Wages were paid in accordance with work completed and time was of the essence. As a result, short cuts were always going to be taken with safety being the first casualty. The at risk behaviour cards came pouring in from every member of the rig's core crew. They showed some very serious breaches of our safety system and head office was informed. What followed would have amazed even the most hardened safety preacher. All core staff were called in, issued with safety tabards and instructed to stop their own work and focus exclusively on policing the dock employees.

The cards came in even faster and jobs were stopped left, right and centre. The dock was falling behind in production and this in turn placed our own rig in jeopardy of losing its charter. Still, the head office maintained the rig would sail early or late, charter or no charter, but most importantly with nobody hurt. We policed the work place for another three days until incoming cards decreased, thus allowing all core crew to return to their own duties and task. However, this lasted for just two days until the number of cards increased again and then the inevitable happened. No, fortunately, not an accident.

Following two high potential near misses involving core crew and dock employees, the OIM was left with no alternative but to shut down the entire dry dock operation associated with our rig. That accounted for 352 men in total. This was a brave decision for an OIM who had just taken command, four months earlier. The OIM spoke through an interpreter to the dock staff, explaining how he did not want to see anybody injured and even said he did not want to make that phone call to their families. People started to take note, especially after he explained that he would continue to stop the job if they

225

continued to be unsafe. After a three hour delay, emails started flying in from senior management from all corners of the world and I was privy to some of the content. After reading these emails I was finally convinced we had what everybody is looking for, namely a safety culture, because the OIM was congratulated for his decision to stop the job by managers in all levels of the organisation.

Imagine the president of a 10,000 strong multinational work force personally thanking an OIM for his decision to stop the job, a job that was now two weeks behind and now many millions over budget. Well, imagine no longer, because that is what happened. The most important point was that the OIM was allowed and encouraged to put the prevention of injury to people first and foremost.

Conclusion

To have an operational safety system following documented procedures is one goal but to have a safety culture is quite another. Having the desire, courage, fortitude and ability to nurture a safety culture is the key to managing not only safety but also the entire business.

Where did I mention the words ISM Code when describing the oil industry in the above text? Not once. Why? Because they never do. They preach and most importantly practice the spirit of the ISM Code through the changed safety culture. There is no need to tell the work force it is a requirement of the Code. By conducting serious training they have adopted a self promoting system which ensures the work force buy into the system and promotes the benefits to new recruits.

It has taken time, effort and money on the rig operator's part to pursue a safety excellence culture. Management have remained loyal and demonstrated their own belief and commitment in the systems. This commitment from the very top that has proved the decisive factor, because they have not just passively agreed to this culture but have been seen by the work force to actively drive and partake in the systems and training.

To prove the success of the oil industry look no further than one of our recent ISM external audits. The same questions were raised by the auditors as I had experienced on board ship, for example "What is the ISM Code?" The big difference is the staff correctly answered all of the auditor's questions with conviction and belief, unaware of the ISM Code guidelines. This is because they are part of a safety culture that sets standards higher than the Code.

Not everything is wrong in the shipping world. There are good professional companies trying and some succeeding with the spirit of the Code. Sea staff are more than willing to adopt and adapt to the future provided they are given the correct lead in adopting to a safety culture and see management believing in it. Many members of ship operators management teams have technical backgrounds and they tend to be driven by equipment efficiency rather than people efficiency. This is not their fault. They have often not been trained in the necessary people skills to achieve this. Safety should not be isolated. It is a part of the whole business picture and thus it cannot be managed well without movers and shakers in the shipping industry having the necessary man management people skills.

The reality at the moment is that like many retro-enforced requirements within our industry, the ISM requirements have just been thrown on board and staff told to read the manuals. You may be able to get around that with a computer but this does not build the required industry culture. Cultures are bred and to breed a culture you need to see a leader. On board it is the master who will lead the crew and encourage, delegate and motivate in the same way as he manages, or should manage, all other shipboard business.

The master needs to be led and his shore management must do this by encouraging, delegating and motivating their senior people. Above all they need to listen to what the master is feeding back. These are the general rules of man management. If the shipping industry safety culture is not working then maybe we should first examine our lack of management skills. We need to address these skills, enhance a culture then introduce a system. If there is no belief you will always have a blind leading the blind culture and our industry will struggle to benefit fromthe positive results of a genuine total safety culture.

In the next chapter Captain Wright builds on what I have set out here and explains how you can chart your own voyage to world class business and safety excellence.

Chapter 13

MAKING ISM WORK IN PRACTICE — THE 'VOYAGE' TO WORLD CLASS BUSINESS AND SAFETY EXCELLENCE

by Captain John Wright FNI

'If you don't know where you're going you will end up somewhere else'

Yogi Berra

Introduction

This chapter builds on chapter 12 and uses the analogy of a 'voyage' to help focus the reader on the desirability of getting to the 'destination' where the management of safety, health, quality and environmental issues has been fully integrated into the fabric of the business. Chapter 14 sets out an example of how one shipping company achieved this. Just like any voyage this one has a starting point, a destination, barriers to getting there (like rough weather or dense fog), a vessel in which to make the voyage, the vitally important 'passage plan' and a cost in time and effort. This chapter describes why the industry and individual companies within it need to make such a voyage, how it can be achieved, the benefits of sailing and who needs to go.

Some of the requirements for the voyage are not pain free because they require business and industry culture change. The UK Confederation of British Industry (CBI) defined culture as "The way we do things around here". This definition reminds us that in some cases 'the way we do things around here' may have to change radically before we achieve world class safety and business excellence. This means putting motivation, involvement, enthusiasm and performance at the top of the organisation's agenda. It means integrating appropriate levels of authority with responsibility and moving the centre of gravity of the business towards the ships.

Culture change also means that some long held industry beliefs, attitudes and behaviours are not appropriate for making ISM successful. This chapter sets out the passage plan, plots the courses and sets out the 'way points' our industry needs to transform our business and enthuse and engage our people in order to make the voyage. This will take inspired leadership from managers with the necessary drive, enthusiasm, determination and clarity of purpose who are uncomfortable with where we are now, can set out a clear vision of where we need to go, have the tenacity to overcome the barriers and then set sail and lead us there.

The process needs to close and seal the yawning gaps between the ships and their management teams. Our industry can no longer afford to have large numbers of its people de-motivated and functioning on half power – if it ever could! Motivation, involvement, enthusiasm and consequential improved productivity from our people are no longer nice to haves, they are essential to business survival in the 21st century.

For the boards of directors, the benefits of the voyage can be real and quantifiable. Business benefits through the use of loss-tracking software can give accountants and finance directors tools capable of quantifying the financial impact upon a company, caused, in part, by inattention to job satisfaction, insufficient soft management skills and the inability to track and eliminate all sources of potential business loss.

Correcting and tracking these things can allow meaningful discussions to take place between owners and their insurers. The aspiration should be to hone the performance and the tools that measure it, to the extent that lower premiums can be negotiated, based on quantified improved risk management performance. A pipe dream do you think? Have a look at these published figures from a leading shipping company:

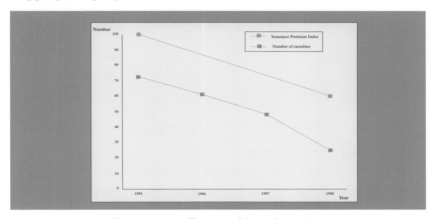

Figure 13.1 — Fleet casualties and premiums

Background

Ship owners have now implemented the ISM Code and STCW '95. Has the change of culture required to achieve effective implementation been achieved and well led? Is our industry getting the maximum benefits from compliance with the new legislation? Are useful and profitable changes to our industry's safety and business culture being brought about, or are we just adding a large number of, often largely unread, manuals to the shelves?

The answers to these questions are to be found in the preceding chapters and results have ranged from spectacular success to dismal failure. What are we trying to achieve? Well, presumably to meet the objectives set out in the Code, which are "to ensure safety at sea, prevention of human injury or loss of life and avoidance of damage to the environment, in particular to the marine environment and property."

This is easily said of course but how can owners give themselves a better chance of meeting these objectives and how can they achieve them cost effectively? The following figure is an overview of the voyage a company can make to achieve real business benefits from compliance with the ISM Code.

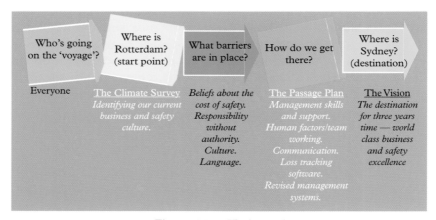

Figure 13.2 — The 'voyage'

Who is going on this 'voyage'?

It really has to be every part of our industry and everyone in it, since it all fits together as one complete whole. However, even on the uneven playing field of today's world of shipping, this book, coupled with my company's own experience, shows us that a quality company can make this voyage alone and there are many business and safety imperatives for doing so.

Where are we now?

Before setting sail the Company must establish which port it is starting from. This may seem obvious but many companies have spent countless hours arguing about where they are now. Why? Because some of them think themselves to be made of chocolate and spend many happy days eating themselves. In other words they feel themselves and their systems to be perfect. On the other hand, some are gloomily convinced they are hopeless in every department. The truth, of course, normally lies in between and must be established by running a strictly confidential company climate survey. The UK Health and Safety Executive (HSE) has just such a statistically validated survey and it can yield some startling results (see figures 13.3, 13.4 and 13.5 below). For the sake of visualisation, let us imagine our departure port is Rotterdam.

The HSE Climate Questionnaire consists of 71 statements and takes about 20 minutes to complete. It has been designed to record people's views within an organisation on ten key aspects of the management of health and safety. The issues examined are recognised as being important in occupational accident and ill-health prevention and managers, supervisors and the work force are asked to express the extent to which they agree or disagree with these statements on a five-point scale. In shipping, the three groups could perhaps be a)shore managers, b)masters and chief engineers (on board managers) and c)the remainder of officers and crew, together with onshore non managerial staff. The questionnaire is purposely designed to seek the views of people in these three discrete groups so that their results can be compared, since there are often wide differences between them. The differing views of sea staff and shore based staff can also be examined, as well as different disciplines. Determining the reasons for these differences will provide important information in seeking routes to improvement. The ten categories are as follows:

1   Organisational commitment and communication.

2   Line management commitment.

3   Supervisor's role.

4   Personal role.

5   Workmates' influence.

6   Competence.

7   Risk taking behaviour.

8   Some obstacles to safe behaviour.

9   Permit to work systems.

10   Reporting accidents and near misses.

Here are some typical questions associated with Factor 6 — Competence:

- *"Am I clear about what my responsibilities are for health and safety?"*
- *"Do I fully understand the health and safety procedures/instructions/rules associated with my job?"*
- *"Do I fully understand the health and safety risks associated with my job?"*
- *"Has the training I had, covered all the health and safety risks with the work for which I am responsible?"*
- *"What action should I take when I am unsure what to do to ensure the health and safety for which I am responsible?"*

Figures 13.3 and 13.4 (from factors 7 and 10 respectively) show results from typical surveys conducted and in both cases suggest that the organisation has severe communication and company culture

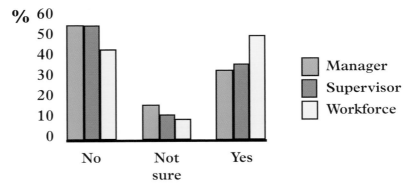

Figure 13.3 — Sometimes it is necessary to take risks to get the job done

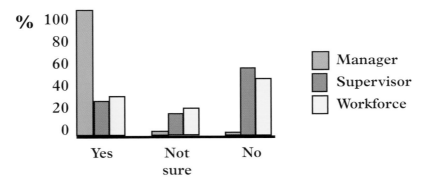

Figure 13.4 — Near misses are always reported here

232

problems. In both cases, especially the second, it appears that the management team are isolated or dislocated from the work force. These results demonstrate that the voyage cannot begin until the management team accept the way they themselves, their supervisors (in this case their captains and chief engineers) and the work force (both afloat and ashore) actually see their company. Then and only then, after accepting once and for all that the starting point for the voyage is 'Rotterdam', however unpalatable that may be, is it possible to make the passage plan to the destination of world-class business and safety excellence.

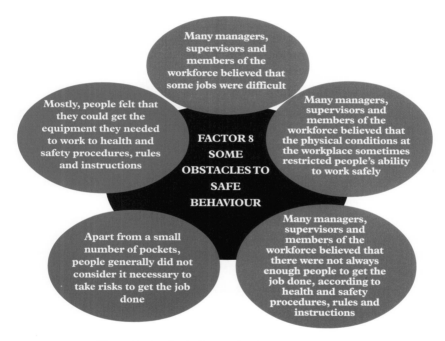

Figure 13.5 — Typical example of results from Factor 8
(red and green indicate negative and positive findings respectively)

The results of the survey must be widely shared with all employees and stakeholders. Communication, ownership and empowerment are the keys to completing the voyage successfully and we shall examine each of these in the 'How do we get there?' section.

Some other useful human factors and leadership related questions that a ship owner or manager might ask during the survey are:

- *"Does a culture exist in our organisation that releases the combined ideas and abilities of the employees ashore and afloat?"*
- *"Are our masters and chief engineers truly empowered?"*
- *"Do our masters and chief engineers have the necessary authority to match their responsibility?"*

If the majority agree or strongly agree with these three questions, then motivation, involvement, enthusiasm and therefore increased safer output will result. Accountability is cheerfully accepted under these circumstances. If the majority disagree, then human factors issues such as poor communication, poor synergy, autocratic management style, excess stress, and punishing and/or ignoring errors and their consequences will all be in evidence. All of these carry a price tag.

- *"Personnel receive the right level of training."*

Agreement with this statement is fundamental to ensuring the risk of errors and their consequences are reduced so far as is reasonably practicable. Training dollars, properly targeted, are a good investment if one examines the costs of accidents that have resulted from a lack of training.

- *"Our seagoing personnel have been directly involved in the preparation of our company procedures."*

If they agree with this statement, you can expect ownership by seagoing staff and that the procedures will be well thumbed, dog-eared and used. You will know they are concise, readable, practical, relevant and available.

- *"We have a genuine fair and just culture."*
- *"All learning opportunity (near miss) incidents are reported, allowing the maximum amount of learning to be extracted without individuals being blamed."*

If they agree with these statements, you can expect to gather in a lot of learning opportunities where there was little or no damage done to people or property. Some of these events will have a high potential to be very costly in lives and/or property but for the intervention of luck. If you know about it you can introduce counter measures (see the 'How do we get there' section). For those without such a culture, the first they may know about a problem is when it has cost a ship, its crew and its cargo!

- *"The designers of bridge and control room layouts on our vessels have taken careful consideration of ergonomics and the human-machine interfaces."*
- *"In taking design decisions, human decision-making, social interaction and potential organisational problems have all been accounted for."*

These things are now taken into careful account in cockpit layouts on aircraft but in our industry we have been slow to recognise their importance. Since all of the above things interact, good equipment design involves far more than hardware engineering. If the application of human factors and ergonomics are an add on and are not integrated with other features of design at an early stage, then we can expect to continue to see too many errors generated. As importantly, the outcome of the errors will be more serious.

Our 'port of destination' (vision for three years hence)

Next, it is necessary to set the Company vision for the port of destination where it would like to arrive in three years time. For the sake of visualisation we shall say this is 'Sydney, Australia'. Why does this come next? Because without this, our ship will steam around in circles outside Rotterdam with no direction, energy or leadership. Why three years? Because it takes this length of time to change the collective attitudes, beliefs and values that are at the core of a company culture and which drive collective behaviours and actions. If you doubt this, ask yourself what the chances are of the attitudes, beliefs and values of an international terrorist changing so radically that he makes the following statement to his followers "*I think that we have been very wrong in the actions we have taken, so I have decided we must surrender*". When you think of any particular infamous terrorist, what do you think the chances are that he will make such a statement? Exactly, so since we have to change attitudes, beliefs and values to succeed, we have no option but to slow steam if we are to have any hope of getting there.

The voyage destination must be set by the board of directors but feel shared and owned by all employees. One way to agree upon the destination is to do so in a series of human factor/team working workshops described in the 'How do we get there?' section. The vision must be Specific, Measurable, Achievable, Realistic and Time-based

(SMART). Typical vision statements defining 'Sydney' can be set around particular questions in the climate survey which would be asked three years hence, such as 85% of all employees (staff and contractors alike) will either agree or strongly agree with the following statements: —

- *"Communication is very effective at all levels, up, down and across this organisation."*

- *"Learning opportunities (near misses) are always reported on our ships."*

- *"There is very efficient and effective feedback on actions taken by the Company as a result of receiving learning opportunities and unsafe acts."*

- *"The senior management team here have safety and welfare of their personnel as their number one priority."*

They can also be framed around accident/incident statistics such as:—

- *"We will have halved our reportable accident rate year on year."*

- *"We will be best in class for our reportable injury rate with a clear gap to the second placed shipping company."*

However, bear in mind that relying too heavily on output indicators such as accident/incident rates can be counter-productive in that personnel may be tempted to cover up incidents in order to ensure achievement of the target. Also remember that the farmer does not fatten his pigs by weighing them!

The above visions are samples only but now you have established your departure and arrival ports you can make your passage plan.

Of course the process of the voyage is about continuous improvement. Thus, upon arrival in Sydney, or preferably before, the team will be planning their next voyage. The journey to excellence never ends.

The vision and your programme for implementation (the 'How do we get there' section below) is a bit like letting a genie out of the bottle. It is exactly the right thing to do and the genie will grant your wishes, if you (the senior management team) are all sincere and genuinely committed. However, if the commitment is insincere and you try pushing the genie back into the bottle once released and letting your peoples expectations down once heightened, then it will be one thousand times worse than not starting the voyage in the first place! Therefore the voyage should include a wider company safety and

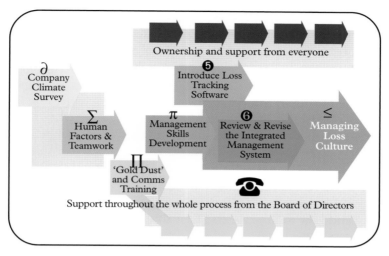

Figure 13.6 — The 'voyage' and 'passage plan' at a glance

business improvement plan, which should have deliverables, milestones and be given the full rigour of project management.

How do we get there?

In this section the passage plan for our voyage from Rotterdam to Sydney is defined. Using the right passage plan starts with understanding these researched facts from the UK Health and Safety Commission:

- Participative decision making, democratic leadership style and individual job satisfaction are all positively correlated with low accident rates.

- While the promotion of job satisfaction is likely to strengthen safety culture, the reverse is also true.

One of the main observations I have made in the years I have had the privilege to talk to and teach several hundred senior officers from many different parts of the world, is the very distinct 'us and them' problem that exists between the ships and their offices. Although there are exceptions to every rule, there is a general dislocation between the two. It is my belief, supported by my own evidence and as highlighted within this book, that this is due to the tendency not to train people who manage men, in the skills of management and communication. Communication in our industry is often of the 'parent/child' and 'my way or the gangway' variety. The on board managers (captains and

237

chief engineers) often do not feel involved and this is evidenced by the fact that they are regularly left out of the decision-making loop.

This problem is less pronounced internally on board ship, mainly due to the fact there are so few people that they tend to work together out of necessity. Nevertheless, I have encountered companies who have gone to a lot of trouble to ensure information arrives on board, only to be thwarted by a master and chief engineer who distribute information on a need to know basis and they appear to believe that their crew do not to need to know! Regularly this is due to a lack of the so-called soft people skills that by and large can be learned but are seldom taught.

Thus this passage plan focuses on people's job satisfaction and welfare and how to improve it. These things impact upon all areas of the business, not just health and safety performance. It is my contention that the industry cannot afford a small army of disillusioned people who are not listened to, are overburdened with paperwork, de-motivated, disenchanted and dissatisfied. Why? Because if they are disillusioned, their performance and output will be severely adversely affected. Industry in general can no longer afford this luxury: Paradoxically, in shipping we have brought this situation about precisely because we have been pursuing efficiencies and have lost sight of the wider business risks involved.

Although using the term health and safety in this chapter, to help focus upon the wider benefits of the ISM Code, the terms 'loss management' or 'loss avoidance' can be helpful. This is because harm to people, vitally important though this is, is only part of the picture and use of the word 'safety' can sometimes 'be a turn off', because it is often wrongly associated with cost rather than benefit. We need to engage the hearts and minds of our accountants, finance directors and the wider shipping insurance and financial sector of our industry and the way to do this is to paint a clear picture of the financial benefits accruing from managing loss. Deciding to manage loss efficiently may well be one of the best business decisions a company will ever make.

The ISM Code is a set of goal setting regulations which require people to use their imaginations as to how best to achieve the objectives set out. Goal setting regulations have proved their value in the oil, chemical and nuclear industries as well as in the criminal and civil law of several countries. This is a marvellous opportunity for the whole industry to come together and examine itself from top to bottom and use this tried and tested route to run the entire industry, rather than to fall back on staid prescriptive regulations.

However, in my experience, people do not take to, nor understand, the benefits of goal setting regulations without guidance from the industry governing bodies (see also Annex 1). In our industry this guidance is rather thin on the ground and without it the chance of success is automatically limited.

The passage plan 'way points' (including departure and arrival ports) are:

1   Conduct a company climate survey (departure port of Rotterdam).

2   Culture change through human factors knowledge and company teamwork.

3   (a) Tapping the knowledge ('gold dust') contained within the work force.

　　(b) Communication skills (engaging, motivating and enthusing people through effective communication).

4   (a) Management skills development (engaging, motivating and enthusing people through effective leadership and management).

　　(b) Personal health and safety commitment to the voyage from the managing director and his team.

5   Tracking, costing and reducing all sources of business loss.

6   Review, revise and implement an effective integrated safety, health, quality and environmental management system including:

　　(a) Risk assessment associated with all aspects of the owner's undertaking.

　　(b) Accident investigation and root cause analysis.

　　(c) Procedures review.

　　(d) Safety management system review and audit.

7   Achievement of a loss management culture (arrival port of Sydney).

The execution of this suggested passage plan will result in increased company and ultimately industry, efficiency and cost benefit through switching the spotlight on to people rather than on to hardware.

This passage plan provides a safe passage to Sydney. Even though there are many course variations you could make for your own company's voyage, I hope that what follows at least stimulates you to do something to ensure you are getting the full potential from your compliance with the ISM Code.

Way Point 2. Culture change through human factors knowledge and company teamwork

Business risk assessment

Human factors knowledge needs to be addressed as part of a company's risk assessment programme. The ISM Code makes risk assessment an implicit requirement and if risk assessment were to be applied from the top of the business to the ships, ports and industry infrastructure, then much industry practice and problem areas would be forced under the spotlight. For example, manning levels, training, fatigue, stress, culture and language, maintenance standards, etc. Figures 13.7 and 13.8 set out the holistic approach to business risk assessment and how it can be managed.

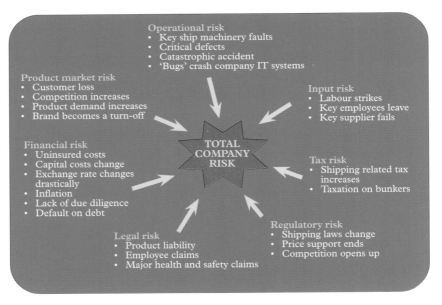

Figure 13.7 — Holistic approach to business risk

Company and industry culture

It is becoming accepted that behaviour is the final common pathway to nearly all loss events and injuries. However, individual and collective

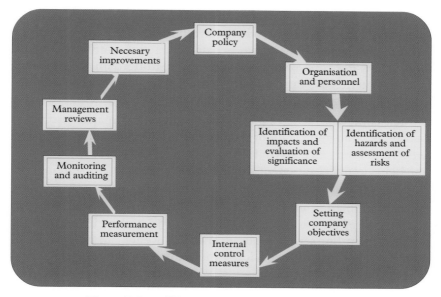

Figure 13.8 — All business risk can be managed with ...

behaviours are driven by our attitudes, beliefs and values. These are the ingredients of company and industry culture and we therefore need to focus on the drivers in the hidden part of the iceberg (figure 13.9) and challenge all of our attitudes, beliefs and values, from the boardroom to each crew member, if we are to make this voyage.

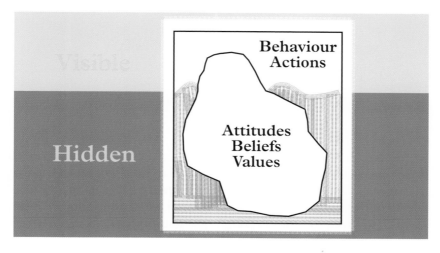

Figure 13.9 — Ingredients of individual and company culture

241

It is the management team who set the Company culture and so the voyage must begin with their commitment. They will have to be the first to challenge their own behaviours, actions, attitudes, beliefs and values if a rope is ever to be let go in Rotterdam. Way point 4 (management skills development), sets out how this can be done by carrying out a management behaviour analysis.

The companies achieving world class safety performance (estimated at a reportable injury rate of 0·02 per 100,000 man hours or better) who have long since arrived in Sydney, invariably focus strongly on the behavioural component in their safety management system.

In so far as the behaviours on our ships are concerned, companies in other industries have realised that it is only by paying attention to the myriad of unsafe acts taking place every day that the occurrence of infrequent but serious injuries and loss events can be prevented. How? By changing the behaviours and the drivers of those behaviours.

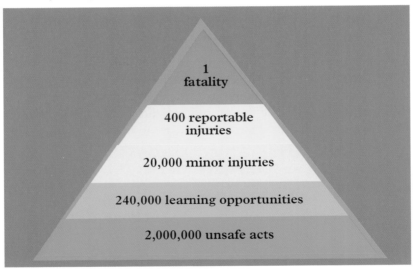

Figure 13.10 — Accident triangle. Data sources: Heinrich, Bird, HSE, John Ormond

Figure 13.10 shows the relationship between unsafe acts and learning opportunities at one end of the scale and serious accidents at the other. So where should we focus our attention? Yes, at the bottom of course, but to get amongst these thousands of learning opportunities we need to foster the correct company and industry culture and steer away from our industry's and the public's tendency to say *"give me a name"* after a loss event has been suffered. We need an effective industry confidential error reporting system and the recently launched marine version of Confidential Human Factors Incident Reporting Programme (CHIRP) which is already used to such excellent effect in aviation, will be a very welcome method of examining our industry's learning opportunities more thoroughly, albeit that the scheme is only operating in the UK at the beginning.

Example of successful culture change

A petrochemical plant engineer was working on critical plant during a shut down and he made a potentially catastrophic error, which he spotted and corrected. He immediately submitted an Opportunity for Safety Improvement (OSI) card to the plant manager. The manager's reaction was to thank him and ensure all the learning was done and root causes examined. No blame was attached, since it is well understood in this company that violations of safety rules with malicious intent are extremely rare unless you are a member of a terrorist organisation! The report received the OSI of the month

The Evidence of a Struggling Safety Management System and Culture
Injury rates remain stubbornly high despite safety procedures being in place
Accidents involving staff, contractors or visitors where 'human error' is given as a cause
Occupational health reports of mental or physical ill-health
High absenteeism rates
High sickness rates
High staff turnover levels
Low level of, or changes, in compliance with health and safety rules
Behaviour or performance issues identified in risk assessments
Complaints from staff about working conditions or job design
Many people believe that the organisation values output above the health and safety of its workers

Figure 13.11 — Evidence of a struggling safety management system and culture

financial award, whereby the local children's charity benefited from a contribution.

What was the reaction to this outcome by personnel in the rest of the plant? Yes, of course, it encouraged the reporting of many more free learning opportunities. The plant had previously never achieved a year without a lost time accident and then achieved three years as the culture changed. It can be done. On the other hand, what would have been the consequences of blaming the individual? (see figure 13.11).

Human factors and team working

We have seen that execution of the passage plan requires industry and company culture change. A very important part of this culture change is gaining knowledge of how humans work and think. This is commonly known as human factors or Crew Resource Management (CRM). I prefer this term rather than the more commonly used bridge resource management because the knowledge and its application extend off the bridge, into the engine control room, throughout the crew and back to the corporate boardroom. This is well understood in the airline business where CRM training is compulsory for all flight deck personnel and the C in CRM stands for corporate as well as crew. In that industry it has been recognised that efforts towards improving safety management, (especially resource management) if they are to be successful, must be driven from the top and extend to all facets of the business.

It is for these reasons that companies derive the maximum benefit from CRM training when the onshore ship managers attend with their captains and chief engineers. Barriers to communication fall, problems that seemed insurmountable evaporate and synergy and teamwork are greatly enhanced throughout the Company. Good leaders do not fear this process; on the contrary, they go out of their way to make sure it happens. This is because good leaders are ready to admit to being wrong and are comfortable to utter the words "*I didn't know that*". They understand that subordinates view these admissions of fallibility as strengths not weaknesses.

The CRM course ensures that vessel managers (ashore and afloat) understand the meaning of, significance and how to improve:

* Synergy and teamwork (on board and organisational).

* Management and leadership styles (on board and organisational).

- Error management and accident causation including root cause analysis.
- Man's mental abilities and limitations i.e. how we deal with our limited mental resources, especially under the pressure of real time.
- Communication, including language and national cultural differences.
- Behaviour types, our ability to pay attention and powers of reasoning.
- Knowledge, experience, workload, competence and confidence and their effects on our judgement and decision-making.
- Stress and fatigue and their effects on the human cognitive (mental) system.
- The effects of new technology on the ship.

Modifying attitudes and thus behaviours through this type of training can significantly improve the output of individuals and the teams they work within. They work more efficiently, tend to make fewer errors and they and the Company suffer reduced consequences of the inevitable errors that do take place. In other words, available resources are managed better and accident rates and consequential costs are reduced.

As has been realised in the airline business, the real mileage in human factors and teamwork training is to be made when it is an integral part of a safety and business management philosophy. Nissan, the car makers, are an example of a company who have just such a management culture. They treat an accident as a production error and a failure in quality assurance. Thus, they very effectively distance themselves from the unhelpful concept of blame.

Does human factors /team working (CRM) training work? Is it worth the effort? This is a note from a ship's master to his manager after his company received written praise for his bridge team working. Ironically, the praise came from the Sydney Pilotage Authority:

"My procedures are based very much on team work. This was one of the principals taught on the crew resource management course. When I did the course I mentioned that it was the best course I had ever been on. I have had no cause to change my mind. The pilot is integrated into the team and is encouraged to share his knowledge with us and to be comfortable about expressing his concerns with anything the rest of the team are doing. Everyone has the right to criticise. This takes some time initially but once the team are confident it will not result in rebuke they are very quick to point out any concerns. One of the most important principles is that we

use the same system at all times. Each port is treated exactly the same. The pilot would have seen the same procedures on departure had he been on board".

To reduce all forms of loss in the fleet and office, the whole organisation needs to be involved. If you apply the principles of the above example across the entire organisation, experience shows us that it is possible to revolutionise an organisation's safety culture while at the same time enhancing efficiency. Not only does the safety culture improve, the people become significantly more committed to their tasks at hand. The reason for this is that a safety excellence culture is really only one facet of a business excellence culture. An organisation which truly values its people will find that all aspects of its business performance will improve. Teaching people to communicate and consult in a caring and effective way will pay great dividends.

Traditional attempts by industry to promote safe behaviour by safety rules and regulations have proved to be effective only up to a point. Whilst rules and regulations are vital components of safety, they can be ignored and usually require vigorous policing. Sadly, some people even adopt the belief that rules are there to protect backs or are made to be broken.

It is for this reason that human factors and team working needs to include an element where the knowledge held within the employees is tapped. This closely involves the people who are likely to be the ones suffering the collision, grounding, fire or personal injury. This important theme is described in the next section.

Way Point 3(a): Tapping the knowledge ('gold dust') contained within the work force

This section sets out one proven method of tapping into the gold dust. The focus is placed upon issues that demonstrate the management team's genuine caring for their people's safety, welfare and job satisfaction and has to be handled with care if it is to achieve the desired outcome. This yields whole company business improvement, not just in safety.

Safety through satisfaction (STS) workshops for seafarers

These workshops can be conducted by trained facilitators at a location to suit the particular Company: They can also be incorporated into the human factors (CRM) training described in the previous section.

It should be borne in mind that front line managers such as captains and chief engineers often suffer from the sandwich filling syndrome;

246

that is, they can feel squeezed from above and below. Their job satisfaction and buying into the culture change process is essential to the success of the voyage.

The objectives of half-day STS workshops are to:

- Extract the large volume of stored knowledge held within your people. This is referred to as gold dust because in terms of the voyage it is so powerful that it can provide the bunkers for the entire trip. It is in this session you find out what is wrong, what is hindering job satisfaction, what is adversely affecting safety and what are the solutions? This comes from the men and women with all the answers — your people.

- Obtain volunteers to be trained in, for example, risk assessment, learning opportunities and root cause analysis. This will create safety and loss prevention champions within the ships and departments they control. In other words, they will be more closely involved with managing their own ships and thus fully bought into the culture change process.

- Discuss areas of concern brought up, such as issues involving communication, risk assessment, training, accident and incident reporting, personal protective equipment (PPE), maintenance and procedures.

After collating the information from the workshops a pattern will emerge and the result is that one can:

1   Present the collated gold dust and solutions to the attendees for their full agreement.

2   Ask the question: What controls, support and resources do you require from your managers?

This information is presented to senior management and the system needed for measurement and controls is agreed. All issues should be risk assessed and addressed accordingly. Issues will have time lines and accountabilities. Some issues will require close examination of the Company management systems but this is always a productive process and should not be feared.

This is the time the necessary authorities are super-glued to responsibilities and in some cases moved on board ship. Accountability for implementing many of the solutions then lies with the seafarers themselves. This they cheerfully accept, providing resources and training needs are synchronised, since it was they who came up with the solutions and people are mightily committed to their own ideas. My experience is that this is a very cost effective exercise and tackles safety improvement at its roots, because the seafarers become part of the solution not the source of the problem.

In short the objectives are to:

- Extract the learning.
- Get full buy-in from the work force, captains, chief engineers and front line managers.
- Decide the controls and resources required.
- Begin the culture change.

The reader will note how closely this model fits with the actions taken by the Company described in chapter 14.

One suggested improvement emerging from this process is the simplification of incident/hazard/learning opportunity (near miss) reporting into one procedure — Safe Learning Events (SLE). The information collected will go into the loss tracking system (Way Point 5) and will be measured in ships, offices and departments at agreed intervals. The SLE inputs will be just some of many inputs into the loss tracking system. This gives the Company a tool with which ships and departments can check each other's loss management performance and create friendly inter team rivalry. It must not of course be used as a stick for managers to beat people with, or it will destroy the trust upon which the new culture is built.

Way Point 3 (b): Communication skills (engaging, enthusing and motivating people through effective communication)

It may seem obvious but communication skills are essential to the execution of the voyage. Without these tools it is impossible to engage, enthuse and motivate people to make the voyage with you. My experience shows these skills, along with the rest of the family of people skills, are often in short supply and need to be learned.

It is imperative that people learn to talk in an 'adult to adult' manner and avoid the 'my way or the gangway' inference if they are to be equipped to seek out sources of loss and reduced job satisfaction. Improving communication skills has to be achieved by discovery learning because it is essential that people be allowed to hold up a mirror to their own performance and self appraise their communication abilities. They are then amenable to being coached in methods of improvement.

248

Deliverables of communication skills training should be:

- To appraise staff of their communication skills abilities (a startling experience for some) and give them tools with which to improve them.

- To develop the communication skills needed to effectively lead, delegate, motivate, make decisions and manage a changing situation.

- To change the way people talk to each other in order to seek out the following:

  - The causes of any loss of job satisfaction.
  - How the causes manifest themselves.
  - The consequences to the efficient running of the job.
  - Why they happen.
  - What will fix the problem.

This training can work hand in glove with the STS programme in order for measured accountability and authority to be placed at, or as close as possible, to the front line of the business.

Way Point 4 (a): Management skills development (engaging, motivating and enthusing people through effective leadership and management)

This involves developing the management skills for middle and senior management. Once again, this can appear obvious. However, it is extremely common to promote people into managerial roles for which they have been given no preparation or training and expect them to pick up the necessary skills by osmosis! The management skills development must:

- Change the way managers think about themselves, their overall lives, their people and the job. They should ask themselves questions such as, "Have I got my business and private life in balance? Do I suffer from excess stress?"

- Allow managers to examine and adjust their behaviours that may be adversely affecting their own, their colleagues and their work force's job satisfaction. This can be achieved by conducting a management behaviour analysis (see below). Another idea is to coach and assist the middle and senior management team to steer the Company vessel efficiently into Sydney harbour. Note that this time this is in its literal sense, since a full scope ship's bridge or engine room simulator can be used for the exercises. The delegates experience management problems in dynamic real time situations, allowing for the coaching and enhancing of managers' abilities and providing them with feedback.

- Change the way their own people look at them. One important way to achieve this is to define, agree and set out a development programme for all staff over a one to two year period. This demonstrates that the management team care for their people and wish to develop them. This development will cover such things as teamwork, leadership, planning, communication, decision making and error management. Another way is to decisively act upon the gold dust and solutions obtained on the STS workshops. The following questions should be answered:

1 What controls and measures do the management team require in the loss tracking system under development?

2 What alarm indicators are required within the programme?

3 What type of measurement and controls are needed?

Management behaviour analysis (MBA)

In syndicate session during the management development training, delegates are asked to:

- Look at the various behaviours that may be adversely impacting both their own and their people's job satisfaction. These will be looked at in the context of an Antecedent (Trigger), Behaviour and Consequence (ABC) analysis:

- List the behaviours and then think through what has triggered those behaviours and the consequences flowing from them.

- Decide the changes needed to alter the triggers and consequences so as to modify behaviours.

- Think out how these changes can be put in place and list them.

- Obtain agreement, if possible, for ABC changes with senior managers. Then syndicates will need to agree measures, accountabilities, time lines and action parties for converting the agreed changes into reality.

- Agree time lines for all behaviours that need more time to consider the ABC changes required.

250

Way Point 4 (b): Demonstration of personal health and safety commitment to the voyage from the managing director and his team

Some 93% of communication is achieved by body language (55%) and the way we speak (38%). Therefore, it is impossible for a manager to disguise the true intent of his message to his people, whatever the words (7%) he speaks, such as, "I put the health safety and welfare of my people above all else". If he does not mean this sincerely it will be seen through in a second!

How, therefore, can the required level of sincerity and commitment be achieved by the senior management team? I have run many workshops where the caring side of people is fully engaged and the moral and legal imperatives of managing health and safety are taken to heart. However, this is not enough because when they return to the daily grind, the old behavioural drivers return to change the message back to 'production is King'. Thus the only sustainable method is to show in crystal clarity the hidden business benefits of a caring senior team who genuinely put the health and welfare of their people first. Points for consideration:

- Look at on board management and budget control. Many safety and business ideas are waiting to be released from your newly empowered teams. Once you do this and show the cynics the evidence of actual dollars saved or gained, it will bring even the hard-nosed to the party! Good safety is good business.
- Peter Drucker (famous business guru) said: "The first law of business is not to make a profit; it is to avoid making a loss."
- All accidents are losses and high potential learning opportunities are big losses, almost suffered but for lady luck.
- Much of the hidden benefits from this voyage will be that the dripping taps of loss will be screwed tight.
- Because this all takes a great deal of effort, convincing the top team of the business benefits is the only way you can sustain momentum. The reason is that for at least some of the time the voyage will require investment and your senior team will discovery learn that investing in health and safety is the most productive thing they will ever do — this is the secret.
- After arrival at Way Point 2 (described above) look to set up small empowered and trained teams on your ships. For example, have them collect their own learning opportunities, investigate

251

their own accidents, come up with their own corrective actions and be accountable for their implementation. This will require training in certain skills but will ensure absolute ownership of not just the problems, but also the solutions to the problems on each ship. I know no more powerful a way to tackle the old 'us and them' problem at its roots. Suddenly we all become 'us' and there is no 'them'.

- Get rid of any petty differences between contractors working long term for you and your direct employees carrying out the exact same task alongside them. Humans have an extremely well honed sense of fair play and these petty things disproportionately affect morale. Individual motivation and therefore performance suffers and this carries a quantifiable cost penalty to the business. Thus any costs involved in levelling out these differences will be paid for by performance.

- The requirement to listen and talk to people will offer untold business improvement opportunities. Why? Because your people will open like roses when the culture changes and they are encouraged to participate fully on the voyage. Out will spill some amazing ideas to improve the business, not only safety, because business improvement and safety are joined at the hip — they are inseparable. They are two sides of the same coin. You will preside over a work force of extremely well motivated people and the advantages to the business are almost limitless.

- Look very closely at the gold dust. Once the top team get this dripping tap of loss philosophy into their heads, they will spot untold business benefits hidden within the material. This is why I christened it gold dust in the first place.

- When reviewing this to do list remember that if you ever say that you cannot afford to do something, always ask yourself if you can afford not to.

- Once some of this is done and the benefits realised, the new-found learning should be taken into IMO. The truth is that the whole industry needs to adopt this philosophy and one or two pioneer companies could bask in the glory of being the visionaries who led the industry on its collective voyage.

Here, to stimulate your own ideas, I have offered a version of commitment to the voyage that could be issued by the managing director to all employees after completion of the STS workshops:

"The purpose of this personal message is to emphasise to you my own commitment to the health and safety of all of our people. It is my number one priority. I fully understand these are just words and you will be able to judge me by my actions. For this reason, I am publishing here my 2004 action plan by which you are invited to judge the genuineness of my commitment in about twelve months from now. In case of doubt, here are my views:

Under no circumstances can I live with having to tell one of our people's partners or spouses that their loved one has been injured, or indeed killed, at work. This is totally abhorrent to me and I am determined to do absolutely everything in my power to avoid ever having to undertake that awful duty. Safety and service are inseparable. You cannot have one without the other. Our company is in the business of SAFESERVICE. No space, no slash, no hyphen! Investing in health and safety is the most productive thing we will ever do. Safety and business improvement are joined at the hip — they are inseparable. They are like Siamese twins.

We have some way to go on the voyage to our vision for 2006 (see the Our Port of Destination section for examples). Since joining the Anon Shipping Company, I have been very impressed with the progress made in our pursuit of safety through the ISM Code. I congratulate all of you on what has been achieved so far but ask you to buckle up for the rest of the trip — we have a long way to go to Sydney!

Tom Peters, the famous business guru says:

• *"You are what you spend your time on. You're as committed — or as uncommitted — as your diary says you are."*

The UK business troubleshooter, Sir John Harvey-Jones says:

• *"All of us who try to change our organisations know that the starting point is to change oneself."*

These are the maxims upon which my personal plan of action is founded. You are invited to judge me and the team ashore upon these yardsticks and upon the plan set out below. Our voyage to world-class business and safety excellence will not founder because the captain and crew were not at their posts!

My plan of action

1 *All senior managers, starting with myself, will be given communications skills training. I personally will carry out not less than one Job Satisfaction through Communication discussion per week with one of you.*

2 *I will discuss health, safety and loss management issues with enthusiasm at the beginning of all meetings in which I participate. Health and safety will be the first thing I talk to you about when we meet. I invite you to correct me if I err.*

3 *I will put in place an action plan for all of the gold dust, safety and business improvement ideas you have given us during the STS workshops and feedback results.*

4   *I will put time and resources into continuing to create a culture where we can more efficiently gather in the near miss and unsafe act data. Near misses will now be renamed learning opportunities. I will feedback progress on this to you.*

5   *In connection with 4 above, we will develop jointly with you the guidelines for a fair and just culture. We recognise that it is very rare that malice is involved in incidents and accidents and so we will together clearly define the way this culture will operate. Our aim is to gather in the hundreds of human factor learning opportunities and put in place corrective actions from these free lessons. Our vision is not achievable without this open culture and for this I need everyone's help. Once again I will feedback progress to you.*

6   *We will create a focus team for each part of the business to help drive our voyage to safety excellence forward. Each team will have clear objectives, amongst which will be the defining of our most important business and safety behaviours, which will be vessel specific and identified and owned by you. These will be published and progress towards achieving routine compliance with them will be monitored by you, the same people who identified them.*

7   *I will review the important root cause issues that arise from some of our Job Satisfaction through Communication discussions and learning opportunities and feed back the results to you.*

8   *We will undertake human factors (CRM) training throughout Anon Shipping Limited. The twin objectives of these workshops will be to improve our team working and to seek solutions from you for some of our remaining areas of concern. Your ideas and any changes implemented will be fed back to you."*

Way Points 1 to 4 achieve the following:

Real and perceived gaps between the ships and management team will have started to close. The senior management team will have got buy in from everyone for controlling and supporting each other. People will have the tools to talk effectively to one another through communication and management skills training. The necessary authorities will have been permanently attached to responsibilities and accountabilities cheerfully accepted.

Way Point 5: Tracking, costing and reducing all sources of business loss

Using a loss tracking software programme will make transparent the fact that productivity and profitability increase within a better safety culture. Every lost-time injury has a human cost, reaching beyond the injured person to the fellow workers close to the accident scene as well as to the family of the injured person. Quite apart from the moral dimension of ensuring employee wellbeing at work, good safety makes good business sense.

One company saved US$2·25 million on insurance premiums alone by managing health and safety effectively over five years. Given a conservative ratio of insured loss to uninsured loss of 1:10, the actual saving may be as much as US$22·5 million. But, of course, the real prize is the 500 serious injuries that were prevented during that period and since then.

Research shows that the average over three day lost time accident costs a company some US$80,000. This money comes directly from the bottom line and since most companies earn at best 10% on gross turnover, a simple calculation shows they have to earn US$800,000 to cover the cost of that single serious accident. Thus, in the course of a year the lost time accidents in a company with a poor safety culture can represent the entire profit margin of the operation.

The costs of accidents and ill health are seriously underestimated. In a 1991 study of a range of industries (the offshore oil industry was closest to the marine industry) it was found that companies in the study were suffering between US$8 and US$36 of uninsured losses for every US$1 spent on premiums for insured losses. Accidents cost one organisation as much as 37% of its annualised profits and another the equivalent of 8.5% of tender price and a third organisation 5% of its running costs. In the case of the oil platform, costs were equivalent to shutting down oil production one day every week! (source: UK Health and Safety Executive).

Operational managers are continuously under pressure to increase productivity levels and become more profitable. They can find it hard to give priority to health and safety issues and consequently people continue to work unsafely. An organisation in this situation finds itself chasing its own tail in a self-defeating, vicious circle of accelerated stress and pressure, spiralling costs caused through work stoppages, more lost time accidents, incident investigations and finally poor interdepartmental and ship/shore relationships. With the monumental stresses upon modern shipping today, the temptation is to accept injury as part of the cost of attaining lower cost bases. This fatalistic attitude will greatly weaken any organisation no matter how confident of success they may be.

The use of loss-tracking software will begin to break down beliefs such as super safety is just too expensive for us. It will do this by:

- Tracking all non-conformances of measures generated during the STS workshops and agreed by all parties.
- Tracking all Health, Safety, Quality & Environmental (HSQ&E) inputs that actually and potentially generate losses. The loss tracking software can be used to cost both actual and potential loss for each entry.
- Enable accident, incident, hazard, near miss and unsafe act reporting to be simplified into one procedure. These Safe Learning Events (SLE) can be tracked by potential as well as actual outcome. Extending cost calculations to these things makes it easier to quantify and then justify the amount of time, money and effort needed to reduce identified risks so far as reasonably practicable. (Note: reasonable practicability is well understood and defined in UK criminal and civil health and safety law.)

Because of the cost implications of loss management, we envisage the management team will treat loss management with equal concern to production issues, without relying solely on the legal and moral imperative and the collective corporate conscience associated with hurting someone.

This new ground breaking software can seamlessly couple the management of HSQ&E and loss with all aspects of the business. Because it can bring the management of loss within the scope and understanding of the finance director and company accountant, it can change the business culture. Sydney will begin to become an achievable destination.

Way Point 6: review, revise and implement an effective integrated safety, health, quality and environmental management system

To describe this part of the voyage could easily require an entire book. Thus, in so far as this description is concerned, I shall point out some of the main areas deserving of your close attention.

From the findings of the voyage to this point, carry out a review of the existing management systems and then revise and implement an integrated safety, health, quality and environmental management system that serves the purposes of the voyage.

Your safety management system should have the type of elements shown in figure 13.12 (overleaf) and each should interrelate as shown.

Your SMS should function as per figure 13.13 (taken from the UK HSE publication *Effective Health & Safety Management*).

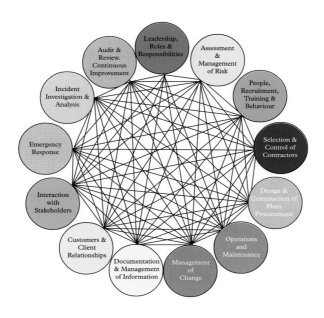

Figure 13.12 — Typical safety management system elements

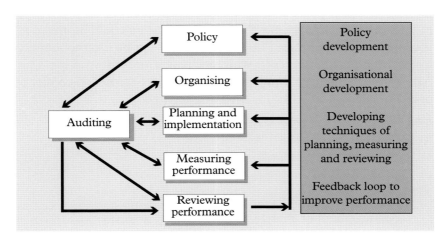

Figure 13.13 — Safety management systems

Any new or revised documents would, of course, need to meet the requirements of the classification society and certifying authority third parties. The revised system should be based on flowtext process maps/flow diagrams and hyperlinks between relevant parts of the system. Hyperlinking is a tool available within Microsoft XP and other operating systems. The advantage of using hyperlinks is that documents or proformas, cross-referenced in the text, can be accessed by clicking on the reference in the text. However, to enable the system to work on board a ship, the ship must be able to access the main server in the office on shore.

The hyperlinks should link one part of the system with another to ensure the system is user-friendly. They should also link the management system to the specific requirements of ISM Code - 2002 edition, ISO 9001:2000, ISO 14001, OHSAS 18001, SOLAS, MARPOL, OCIMF and all other relevant marine regulations, codes and guidelines. The revised system will therefore, transparently show how a ship owner or manager meets all regulatory requirements, codes and guidance.

Another tip in revising your safety, health, quality and environmental management system is to follow the principles set out in the UK Health & Safety Executive's publication *Successful Health & Safety Management* (HSG 65). This will ensure the system is at the leading edge of SHQ&E management systems thinking.

Implement an electronic audit database that will record all audit findings, produce audit deficiency reports and produce trend analyses in the form of pie charts, etc. The proposed system could be linked to the loss tracking software (Way Point 5) and can be used to identify lost time and associated costs with mechanical breakdowns, etc. This type of system has been developed in a marine / construction company over a number of years and is tried and tested.

Essential elements in a typical SMS are:
- Leadership and organisation.
- Performance standards.
- Recruitment and training.
- Communications.
- Management of change.
- Accident investigation and analysis.
- Selection and control of contractors.

- Purchasing controls.
- Monitoring, inspection and auditing.

Leadership and organisation requires:
- Policies.
- Commitment.
- Responsibility.
- Accountability.
- Authority.
- Motivation.
- Safety organisation.
- Planned programmes.
- Goals and objectives.

The required elements for performance standards are:
- Who does what, when, how and to what effect.
- Identification and assessment of all potential hazards and risks
- Procedures for safe systems of work to control risks associated with normal and abnormal operating conditions, plant and equipment.
- Emergency procedures.

Recruitment and training have to be considered for:
- Managers.
- Employees.
- Contractors' employees.
- Identification and provision of relevant skills and competencies.

Communication is required:
- Downwards.
- Upwards.
- Laterally between individuals.
- Between work groups and teams.
- Safety representatives and committees.

Management of change is needed in relation to:
- Plant and equipment.
- Processes.
- Personnel/contractors.
- Procedures.

The required elements for accident investigation and analysis are:

- Investigation of minor accidents and learning opportunity incidents as well as serious ones.
- Identification of underlying causes as well as immediate ones.
- Procedures to act upon findings and prevent recurrences.

The required elements for selection and control of contractors are:
- Pre-contract assessment.
- Monitoring during contract.
- Post-contract assessment.

The required elements for purchasing controls are:
- Equipment.
- Materials.
- Services.

Monitoring, inspection and auditing:
- Closing the loop in all areas of the SMS.

Way Point 6 (a) Risk assessment associated with all aspects of the owner's undertaking

Again, this is a vast subject in its own right. Here, I have described some of what I consider to be the most helpful aspects. Figure 13.14 below shows one proven method to brainstorm the hazards associated with a particular task. It is called a cause and effect analysis, or fishbone

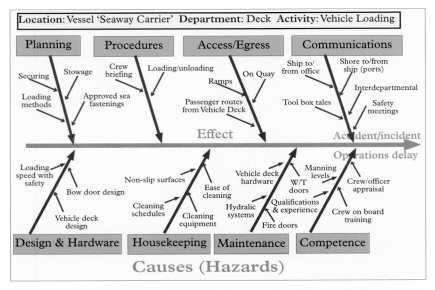

Figure 13.14 — Risk assessment for Ro-Ro ferry operations (cause and effect analysis)

diagram. Assemble a risk assessment team who are closely related with the task in question and build the fishbone on a whiteboard. This method works well because it uses the right brain, or picturing brain, to build up the picture of the hazards involved. A list of hazard guide words helps reduce the chance of oversights.

Once you have uncovered all the hazards and the risks associated with those hazards, you now need to assess the risks. The risk assessment matrix (figure 13.15) will assist you in this task:

For each risk the likelihood of occurrence and severity is assessed using a simple 1 to 5 scale for each. The risk factor is the likelihood multiplied by the severity.

		Severity				
		1 negligible injury, no absence from work	**2** minor injury requiring first aid treatment	**3** injury leading to a lost time accident	**4** involving a single death or serious injury	**5** multiple deaths
Likelihood of occurrence	**1** a freak combination of factors would be required for an incident to result	1	2	3	4	5
	2 a rare combination of factors would be required for an incident to result	2	4	6	8	10
	3 could happen when additional factors are present, otherwise unlikely to occur	3	6	9	12	15
	4 not certain to happen but an additional factor may result in an incident	4	8	12	16	20
	5 almost inevitable that an incident would result	5	10	15	20	25

Figure 13.15 — Risk assessment matrix

1—6	May be acceptable, however, review task to see if risk can be reduced further.
7—14	Task should only proceed with appropriate management authorisation after consultation with specialist personnel and assessment team. Where possible, the task should be redefined to take account of the hazards involved or the risk should be reduced further prior to task commencement.
15—25	Task must not proceed. It should be redefined or further control measures put in place to reduce risk. The controls should be re-assessed for adequacy prior to task commencement.

Key to figure 13.15 — risk assessment

By redefining the hazard severity, risk evaluation matrices can be used to assess health, production and environmental risk as well as the risk of accident and injury. For example:

1   Negligible injury or health implications, no absence from work. Negligible loss of function/production with no damage to equipment or the environment.

2   Minor injury requiring first-aid treatment or headache, nausea, dizziness, mild rashes. Damage to equipment requiring minor remedial repair, loss of production or impact to the environment.

3   Event leading to a lost time incident or persistent dermatitis, acne or asthma. Localised damage to equipment requiring extensive repair, significant loss of function/production or moderate pollution incurring some restitution costs.

4   Involving a single death or severe injury, poisoning, sensitisation or dangerous infection. Damage to equipment resulting in production shutdown and significant production loss. Severe pollution with short term localised implications incurring significant restitution costs.

5   Multiple deaths, lung diseases, permanent debility or fatality. Major pollution with long term implication and very high restitution costs.

Another tool to assist you is the Pareto Analysis or 80/20 Rule (figure 13.16). Simply stated this means that for any particular subject that you brainstorm, including risk assessment, you will find that 80% of the solutions can be found within 20% of the problem areas.

So in the figure overleaf, if the risks associated with 20% of the areas which contain hazards are reduced so far as reasonably practicable (in this case in the areas of communication and maintenance) you will have reduced 80% of your risks to As Low As Reasonably Practicable (ALARP).

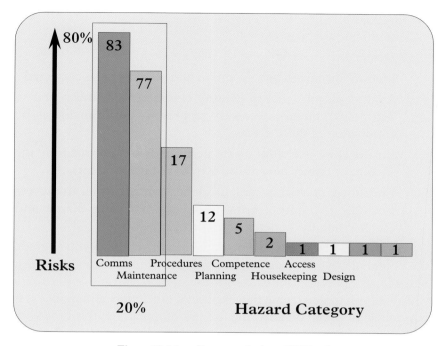

Figure 13.16 — Pareto analysis — 80/20 rule

Way Point 6 (b) Accident investigation and root cause analysis

Once again this is a very large subject. Here I have described some of what I consider to be the most helpful aspects.

The main aim of accident investigation is to prevent the recurrence of similar accidents/incidents and, therefore, the actual causation is a matter of prime importance. The aim of accident investigation is not to lay blame, so why investigate? To:

- Establish all the facts relating to the incident.
- Draw conclusions from the facts.
- Identify the immediate and underlying causes.
- Identify weaknesses in management control.
- Calculate the financial costs to the organisation.
- Ensure that a repetition of the chain of events is not possible.
- Make recommendations to prevent a recurrence of the incident.
- Improve/develop an open culture.
- Share lessons learnt with others.

263

Accident investigation and follow-up

Systems should be in place to:

- Determine the root cause of each incident.

- Identify specific follow up action and systems to be corrected.

- Analyse all incidents to identify common root causes and to determine changes necessary to prevent future incidents by elimination of those causes.

- Ensure close out of follow up items and assess or measure the success or failure of actions taken to reduce incidents.

- Encourage open and frank incident reporting by all employees through reducing emphasis on apportioning blame and emphasising the benefits of lessons learned.

Figure 13.17 is an incident potential matrix. It allows you to grade the potential of your incoming learning opportunities (near misses). Knowing the potential for causing harm of each learning opportunity will allow you to determine the amount of time effort and money it is reasonable to spend on ensuring the risk of repetition, this time with perhaps far more serious consequences, has been reduced so far as reasonably practicable.

For example, the X in box A1 could represent a near miss involving a main engine crank case incident in the engine room of a vessel where no damage was caused or injury sustained but the potential loss if an incident had occurred could be of the order of US$1,000,000 (box D4).

Figure 13.18 is a model for root causes of accidents or latent pathogens (taken from Professor James Reason's research, professor of psychology at Manchester University, England). They are called pathogens because the accident causing factors lie in the system, sometimes for years, until the correct triggering mechanism completes the chain of events leading to an accident. Correction of identified latent pathogens lies within the management system. Accepting the huge deliverable from meaningful root cause analysis in the management of loss is one of the major keys to the successful and profitable completion of this voyage.

Way Point 6 (c) Procedures review

I have described some of what I consider to be the most helpful aspects.

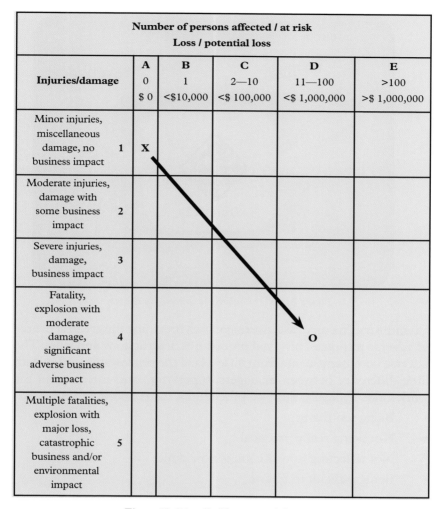

Injuries/damage	Number of persons affected / at risk Loss / potential loss				
	A 0 $ 0	B 1 <$10,000	C 2—10 <$ 100,000	D 11—100 <$ 1,000,000	E >100 >$ 1,000,000
Minor injuries, miscellaneous damage, no business impact 1	X				
Moderate injuries, damage with some business impact 2					
Severe injuries, damage, business impact 3					
Fatality, explosion with moderate damage, significant adverse business impact 4			O		
Multiple fatalities, explosion with major loss, catastrophic business and/or environmental impact 5					

Figure 13.17 — Incident potential matrix

How do employees perceive safety procedures?

The UK Health and Safety Executive Climate Survey Tool as described in the 'Where are we now' section is a useful way of ascertaining the attitude of personnel within an organisation to health and safety procedures. One of the elements, entitled Obstacles to Safe Behaviour, includes a number of statements in relation to procedures, instructions and rules.

Our company has been involved in administering the Climate Survey Tool for various companies involved in the North Sea offshore oil and

265

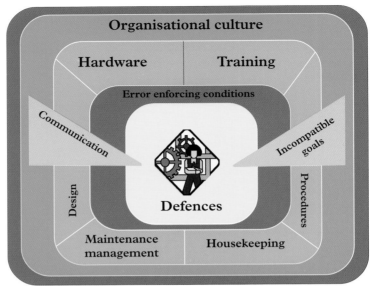

Figure 13.18 — Underlying causes of accidents

gas industry. In every case, the major area providing the greatest degree of adverse responses involved procedures, instructions and rules. The adverse comments come from all levels of the organisation, often with little difference between managers, supervisors and members of the work force. Typically, they see procedures, instructions and rules as:

- Being too many.
- Not being really practical.
- Not reflecting how the job is now done.
- Being difficult to follow.
- Being there to protect the management's back.
- Not being followed.
- Being developed/reviewed without involvement of personnel, especially supervisors and members of the work force.

Whilst the Climate Survey Tool can identify procedures as one of the obstacles to safe behaviour, the statements are so broad that they cannot assist in identifying exactly which procedures, instructions and rules are the real problem. However, by following up these findings during the STS workshops and face to face discussions with on board managers, it is possible to identify the specific procedures with which there is difficulty and to obtain potential solutions from those best

266

placed to provide them, i.e. the people using the procedures. This approach can provide a great deal of buy in from the staff. They immediately feel involved because management is asking them what the problems are and actively seeking their help in solving the problem. This can be very much of a win/win situation for everyone involved and fits with the ethos of the voyage and required culture change.

There are lessons here for the development and implementation of new procedures as well as achieving acceptance and compliance with existing procedures.

What constitutes a good health and safety procedure?

The UK HSE's publication HSG 48 *Reducing Error and Influencing Behaviour* provides some useful guidance in this area.

It should be borne in mind that even if procedures are not always formally written down, they exist through the working practices of the staff. HSG 48 suggests that ideally, procedures need to:

- Be accurate and complete.
- Be clear and concise with an appropriate level of detail.
- Be current and up to date.
- Be supported by training.
- Identify any hazards.
- State necessary precautions for hazards.
- Use familiar language.
- Use consistent terminology.
- Reflect how tasks are actually carried out.
- Promote ownership by the users.
- Be in a suitable format.
- Be accessible.

Way Point 6 (d) Safety management system review and audit

Each organisation must develop its review and audit techniques and they should be tailored for the shipping industry in general and the individual Company in particular, such that the system remains viable in all areas. It must include methods that allow measurement of the system to take place and for ongoing and continual review to become a natural occurrence.

267

To maintain the whole policy and thus the safety management system, a regular audit of what is actually taking place must be conducted. This operation ties up all of the other stages in a single course of action. Audits should never be feared but be used as a means of ensuring all aspects of the SMS are functioning as intended.

Barriers to the voyage

One of the biggest barriers to overcome is the general resistance to change and this has to be managed, both by individual companies and by our industry. This was summarised neatly nearly 500 years ago:

"It should be borne in mind that there is nothing more difficult to handle, more doubtful of success, and more dangerous to carry through than initiating change. The innovator makes enemies of all those who prosper under the old order, and only lukewarm support is forthcoming from those who would prosper under the new. Their support is lukewarm partly from fear of their adversaries, who have the existing laws on their side, and partly because men are generally incredulous, never really trusting new things unless they have tested them by experience. In consequence, whenever those who oppose the changes can do so, they attack vigorously, and the defence made by the others is only lukewarm. So both the innovator and his friends come to grief."

Machiavelli, 'The Prince', 1514

Another barrier is the tendency towards going to the lowest bidder, the folly of which is contained in this quote:

"It's unwise to pay too much, but it is worse to pay too little. When you pay too much, you lose a little money - that is all. When you pay too little, you sometimes lose everything, because the thing you bought was incapable of doing the things it was bought to do. The common law of business balance prohibits paying a little and getting a lot — it can't be done. If you deal with the lowest bidder, it is well to add something for the risk you run. And if you do that, you will have enough to pay for something better."

John Ruskin, 1819—1900

In the last eight years I have taught human factors (CRM), accident investigation and risk assessment to hundreds of mariners and during these sessions we have asked our delegates to tell us the barriers that limit their personal performance at work. We have also asked them how the problems could be rectified. The following is a selection of the most often recurring themes and although the author would not agree with all of their solutions, both the problems and remedies offered are generally of a high quality and offer food for thought. The comments are quoted here verbatim:

• *Excess working hours, causing tiredness.*

Fix: Carry out a manning level review and carry out a higher level of maintenance. Junior engineers need to have gained more experience to allow more delegation and this requires retention of staff and care taken with accelerated promotion.

• *Traditional tendency to autocratic management style at senior levels, which leads to lack of delegation.*

Fix: Expose our people to other examples of how to do things well.

• *Language difficulties on board.*

Fix: Continuity of staff helps. Minimum level of English required to be proved before entry.

• *A lack of the required level of experience in both myself and my officers.*

Fix: Advanced training is required to accommodate the increasing number of new procedures and regulations etc. that are constantly being introduced by the industry in general.

• *Shore pressure in the form of visits by Port State control, surveyors, immigration etc.*

Company fix: Try to reduce the number of port inspections by the staff of different organisations who visit the ship during port turn rounds.

Industry fix led by the Company: There is a need at industry level to standardise this duplication in inspections across the world. The quality and level of training that these inspectors have received needs to be reviewed. All too often inspectors pick up on the minor points to the detriment of the more serious points.

Ship fix: Delegate where possible and improve forward planning, whereby these visits can be so structured and organised that visitors fully understand the importance of the role of the master in carrying out his main function, that being the safe operation of the vessel at all times. Try not to conduct inspections at the same time as other operations are taking place.

• *Excessive reporting from ship to company and vice versa.*

Although all of these reports are being forwarded to the Company, very often concerns of a more serious nature, as detailed in the reports, are being missed. This may well be a function of the sheer volume of paperwork. Therefore, the perception is that this paperwork is detracting from the main role of the master. A lot of these paperwork problems have arisen as a result of the introduction of the ISM Code.

Company fix: Ensure a balance is struck between the requirements of the ISM Code and the practical implementation of same, without vast amounts of paperwork being produced.

- *Lack of time to concentrate on the job in hand.*

    This is due to many other tasks set down for the master. For example, where a report on a certain topic is required to be sent to several different parties, the report forms all differ in their content and this is very time consuming. For example, incident report forms.

    Company fix: Develop and introduce standardised forms. Achieve this by consultation with all interested third parties to try to avoid this unnecessary duplication. Companies and certifying authorities must look at their own forms to reduce any unnecessary duplication.

- *A new employee finds the amount of paperwork excessive.*

    There needs to be a guide for a new employee to follow that will introduce him/her to the best way of accessing and accommodating all the paperwork.

    Fix: Review and standardise paperwork across the fleet to cut down on unnecessary paperwork.

- *Lack of a good on board social structure*

    where cross fertilisation of knowledge experience and information could take place. There is a lack of a social area on some ships to meet and talk, or the space allocated for these important social gatherings are too small and there is a tendency for individuals to disappear to their cabins and not socialise at all. These existing small areas are supposed to support a range of activities such as watching TV, where people can smoke, talk, but cannot physically accommodate the number of personnel on board. There is a need for a good social structure to be implemented with the encouragement of the Company. This increase in communication and social skills would enhance the overall running of the vessel and help reduce stress that some people are tending towards.

    Fix: The Company to provide better facilities to improve interpersonal relations, for example, larger common rooms or smoke rooms.

- *With the very rushed and intense workload*

    around the coast, other important planned maintenance is falling behind. Whilst it is important for those involved in watchkeeping to be well rested to accommodate the schedule, this in turn limits the number of staff to complete work lists and planned maintenance. Superintendents do not entirely understand this problem when planned maintenance falls behind.

Fix: Where necessary, provide additional personnel, over and above the normal ship's compliment, to undertake other necessary work.

• *With the implementation of many check lists and procedures*

I feel that my skills and knowledge are not being challenged and that the checklists/procedures are being used as a substitute for training. Some young officers tend to treat them as such and are not really increasing their knowledge and experience. They just follow forms to complete a task. I am not sure how some would cope in a real emergency or some task presented to them that is out of the norm.

Fix: Ideally it would be beneficial for cadets to be placed on older ships where hands on work still plays more of a part in the overall running of the vessel. This will enhance knowledge and increase their level of competence.

• *Some senior officers, both deck and engine,*

lack basic management skills. Their presence in the Company creates a bad atmosphere for those who have to work with them. Some good officers have left the Company due to the management style of these individuals.

Fix: Introduce an officers' assessment programme whereby all officers are periodically assessed and where necessary, management training can be implemented if it is deemed necessary. Where continuous poor management styles are affecting the running of the ship, warnings should be given in exactly the same way as they are to other officers.

• *Poor morale*

Some people feel they are left in the dark because the flow of information is at times poor. They need to know whether the Company is doing well or not, i.e. are we doing a good job? More discussion over the budget in the sense we masters would like to know how we are performing, budget wise, so we have an opportunity to make improvements.

Fix: Improve communication. Show people are valued by giving praise when deserved.

• *No time to catch up on personal learning preferences*

and having the time to place more of a personal emphasis on

safety inspections and procedures. Excess paperwork in other areas is preventing me from doing this. There is a need for the Company to discuss with bodies like IMO on the amount of paperwork that has been generated in the formulation of things such as the ISM Code and other statutory procedures. A sensible conclusion needs to be reached whereby the ships can be run safely, without excessive amounts of paperwork to achieve this vitally important aspect.

Barriers emerging from the rest of this book

Quote from survey contributor:

"The ISM Code has created a paper chase on board vessels which fails to address the main problem, crew training. Some companies assume that by having an auditable system that complies, is a substitute system for good quality crew. Audits are carried out in accordance to their documented system — ISM — just as with the QA auditors — it doesn't imply its good i.e. because a chocolate factory complies to its QA system it doesn't mean the chocolate tastes good. Similarly, a ISM system on a vessel doesn't mean the vessel is any safer or good, it just complies to a auditable system — does it?"

My comment: The writer makes a fundamental point about the need for continuous improvement. To achieve this we need a shift in the industry culture from top to bottom. Reward the best, punish the worst and construct our industry accordingly.

Quote from survey contributor:

"The concept of ISM is obviously a first class idea. To the very good companies it made no difference as essentially all ISM was doing was formalising what was their normal operational practice. To other companies ISM was perceived as a bureaucratic burden and was operated with reluctance. ISM will only operate well if the owners wish it to. On some vessels the volumes provided by the owners fill whole shelves and are clearly never read or used. A conscientious attempt should be made to limit the number and size of the ISM manuals. The lack of continuity of service by crews who are generally supplied by agencies is not conducive to the efficient operation of ISM. ISM has to be implemented on board as a way of life, but in reality on most vessels it is perceived as yet another paper work burden that further limits the time available to actually operate the vessel. If there is a delay to the vessel or cargo operation due to an ISM defect then it is expected that this would be held against one or more members of the crew and could lead to disciplinary action or dismissal, hence the reluctance to report defects."

My comment: This observation highlights the need for culture change for our entire industry. The voyage should be made by everyone.

Quote from Mr W.A. O'Neil, Secretary General to IMO:

"The ISM Code, by comparison, concentrates on the way shipping companies are run. This is important, because we know that human factors account for most accidents at sea and that many of them can ultimately be traced to management. The Code is helping to raise management standards and practices and thereby reduce accidents and save lives."

My comment: This is an endorsement of the need to make the voyage.

Quote from Swedish Club:

"The three most important factors for a properly functioning safety management system and improved safety records were:

- *Commitment from the top management ashore.*
- *Increased safety awareness on board.*
- *Checklists/procedures for key shipboard operations.*

The reasons for a non-functioning safety management system were:

- *Too much paperwork/documentation.*
- *People do not know how they are expected to use the system.*
- *People do not believe in ISM.*

The top four proposals on how to convert a poorly functioning system into a properly functioning one were:

- *More ISM training and education.*
- *Reduce paperwork/documentation.*
- *Provide seafarers with good examples of ISM in practice.*
- *Make sure that the accident reporting procedures work and increase feedback from accident reports to seafarers."*

My comment: Once again, these findings are directly in tune with the principles set out in the voyage.

Quote from Swedish operations manager:

"High turnover prevents SMS from working efficiently. Training in basics is forever ongoing and is never completed because it has to start all over again. Crews are too small to enable training on SMS and occupational work and do the work properly at the same time. Crew from manning agents do not always take company goals and objectives to their heart because they will be gone after a few trips. There is lack of ownership because of short duration employment. Others with long term employment recognise however the importance of a functional SMS because it makes their job easier by providing routines and the basis upon which to train the newcomers."

My comment: Providing continuity of crews is cost effective in the long run and the industry culture should shift to accommodate this.

Quote from survey contributor:

"All accidents/near misses/dangerous occurrences are reported so that all the other vessels in the fleet can maybe benefit from our experiences and vice versa. The majority of non-conformities can be dealt with on board quite soon and even these are promulgated throughout the fleet."

My comment: Also, people are more willing to discuss the near miss or unsafe act or condition because nobody is carrying the baggage of a victim, which produces the background noise of emotion and the need to find someone to blame.

When a company arrives in Sydney it will be routine for all such incidents to be subjected to a no blame root cause analysis, with all the management system flaws highlighted.

It is extremely rare for a seaman, or anyone else for that matter, to get up in the morning and say *"Aha! — this is the day I have been waiting and planning for — I am going to blow up the main engine today."* This is not normal behaviour and thus the accusing finger and punishment is not normally an appropriate counter measure. Furthermore, it is guaranteed to drive learning opportunities underground. ISM skills training to gain uniformity across the shipping industry should be provided. It should include management training in the soft skills so that one may expect people to gain an understanding of why they need to harness their people power.

Conclusion

What benefits could the industry, ship owners and ship managers expect from completing the voyage. Well they can expect to see:

- Safer ship operations and cleaner seas.
- Avoidance of the pain and suffering caused by employee and contractor accidents.
- Improved quality and productivity as people routinely get it right first time.
- Enhanced performance through more effective engagement and communication.
- Better employee relationships as employees recognise that the Company and its management care for their well being and that of the environment.

- Reducing the occurrence of accidents and incidents.
- Protection of the company's reputation by avoiding adverse publicity and creating a positive image with employees, contractors, and customers.
- Kudos for the Company, from the achievement of an improved safety and business performance.

If the voyage is well led and embarked upon as part of a business improvement drive rather than a bolt on safety initiative, then the following positive outcomes can be expected:

- Communication will improve. It will improve on board, between sea and office staff and up, down and across the Company. The deliverables are many and various and most will positively affect bottom line profits. Shop floor ideas will flow uninterrupted onto the bridge and thence into the boardroom and vice versa. Managers ashore and afloat will 'walk the walk', not just 'talk the talk'. Motivation, involvement, enthusiasm and therefore productivity will increase.
- Power and authority will be effectively delegated to the on board vessel managers (master and chief engineer). This empowerment will ensure that responsibility and accountability is cheerfully accepted. Outcome: increased efficiency.
- Vessel managers and their officers will fully understand the concepts of hazard, risk, accidents, incidents, learning events and loss control theory. This knowledge will ensure that;

  a Accident and incident causation is properly understood (underlying and immediate causes) and that all high potential incidents are subjected to root cause analysis. Far too many incident investigations are ineffectual, shallow affairs that don't even scratch the surface of causation.

  b Accidents and incidents are analysed using their potential rather than actual outcome. The former is arrived at objectively, the latter is a function of lady luck.

  c A fair and just culture is fostered in order to gather in all the learning opportunities. It should be remembered that errors and intelligence are opposite sides of the same coin. We cannot eliminate errors.

- Accidents, incidents and near misses, together with all sources of loss will be measured, recorded and priced, both actual and potential costs. This may be the single most crucial outcome from making this voyage, because it will foster boardroom buy in.

The voyage

There is no doubt that individual companies can do a lot to instigate and execute this passage plan. However, for the ISM Code to achieve the aims of its authors much more needs to be done:

We need trained inspirational leaders who will head the drive to industry culture change. They need to revitalise, enthuse, engage and transform our people and business.

As part of this voyage to world-class business and safety excellence, training can be used to hone leadership skills and help people to understand themselves, their natural human weaknesses and how to effectively guard against them. To be effective it must form part of a commitment to an overall management culture within each company that has, at its core, a commitment to and a trust of its people.

People generally go to their ships and their shipping offices with the intention to do their very best. This is natural. They do not go with the thought of what they can mess up today or what accident they could cause. What normally stops them doing their best is stifling bureaucracy, poor communication up down and across the organisation, lack of authority, being set incompatible goals, trying to follow unworkable procedures and a hundred other related barriers. Work related stress is the most common outcome and that adds to the adverse affects on production as well as safety.

If our industry leaders do decide to set course for Sydney and world class safety and business performance, the biggest winners of all will be the members of the global village, since we shall all have safer ships and cleaner seas.

Note:
"I am most grateful to John Ormond Management Consultants Limited for stimulating conceptual ideas contained within some of the diagrams in this chapter."

Chapter 14

IMPLEMENTING THE ISM CODE IN A MEDIUM SIZED FLEET

by Captain Sean Noonan FNI

"The cornerstone of good safety management is commitment from the top. In matters of safety and pollution prevention it is the commitment, competence, attitudes and motivation of individuals at all levels that determines the end result."

From the IMO's ISM Code preamble

Introduction

There are many different ways in which to write and run a safety management system. This chapter deals with experiences in Tarquin Gas Carriers (TGC).

Part 1 looks at the thoughts behind the system and how the Company used our knowledge and understanding to help us create effective tools for the management of the fleet.

Part 2 looks to the actual implementation of the system and how it was explained to all levels of staff.

Part 3 examines the system of continual improvement and how, once the momentum was started, the Company progressed towards its objective.

Concepts and philosophies — setting the culture and commitment from the top

The goal is the improvement of safety and environmental protection, in order to reduce exposure to human errors, particularly one man errors (*we are all fallible*) for all aspects of the shipping venture.

The Safety Management System (hereafter referred to as SMS) should be nothing particularly new to an experienced seafarer but should formalise the way in which all persons conduct the different jobs throughout the Company so that a consistent way, that is easy to follow and easy to learn, is used and applied throughout the Company. It will allow all staff to input their thoughts on how to improve the processes. We should all learn from each other's experiences, insights and mistakes. Everybody has their own style and different strengths and weaknesses, by pooling our information we can share our strengths and compensate for our weaknesses.

To achieve a SMS that works from the requirements of the ISM Code requires understanding of the concepts and philosophies behind its existence. This understanding of the concepts and philosophies of the ISM Code are very well addressed in the earlier chapters of this publication. It also requires full cooperation and commitment from the owner and top executive directors and managers.

Many people I have met believe the ISM Code and its requirements are an additional burden on the industry, but in reality it is just the formalising of existing best practices and, if set up correctly, won't increase the necessary work load but actually make the working environment much safer and business objectives easier to achieve. I know that if I had had an effective SMS throughout all of my career I would have been more effective than I was in my work and with my progression though the ranks to master.

The foundation of the system has to be in its planning. A solid well-planned foundation will allow the construction of a system that can weather the storms of the unpredictable sea borne venture that is shipping.

Finding and choosing the right people for the tasks of writing the various parts of the SMS is very important, as not only professional experience, but also a sense of the overall working of the system and all its components is needed to appreciate; not just what is written, but also how it will be applied and how the component parts complement each other.

The culture that exists within the Company, both ashore and on board the ships, needs to be prepared and nurtured. The selection of right minded staff is very important. Employing a person who believes in the effectiveness and potential for the ISM will give benefits, reduce the difficulties in implementation and ensure the continued improvement of any system.

Another requirement for the SMS and for the top management to consider is ensuring that adequate resources (personnel and equipment) are in place throughout the Company to allow the processes for management to be effective.

The bonus of an effective system is that it will lead not only to improved safety and environmental protection but also to savings in costs and greater control of the unpredictable events that are outside the standard management budgets that are the bane of the ship owner. However, initial costs will have to be met and the plugging of training gaps can apparently be expensive. As you will see, the costs involved in the setup of a system are investments in the future and mean that in the long run substantial overall savings will be made.

Setting the objectives

The objectives that are set out by the Company should be simple, understandable and realistic. Our objectives were defined in our policies and backed up in the management reviews. Below is a sample of the main objectives that we set in TGC.

C.2.1 Company policy on safety and environmental protection

It is the policy of the Company that in all its activities, the highest priority will be given to the need to protect the health and safety of its employees and other involved persons, and to conserve the environment.

The Company aims to eliminate work related injuries and illnesses, by requiring high standards of working practices, and by providing a safe working environment for all. Identified risks to health, safety and the environment are addressed by the SMS. The Company involves employees in the promotion of these aims, requiring the reporting of previously unidentified risks, and in monitoring of the measures taken to improve the company's SMS.

The Company demands commitment by its employees to its high standards in respect of health, safety and environmental protection, and ensures that they have the skills and support to meet these. It assists and encourages the masters of its vessels in the implementation, maintenance and verification of the SMS on board.

The Company requires observance of all applicable laws, industry codes and appropriate guidelines. It seeks to identify and provide for training needs. This will be achieved by constant monitoring and involvement of all office and sea staff, in continuously improving the safety management skills of personnel, in particular in respect of emergencies related to safety and environmental protection.

This policy must be understood, implemented and maintained by all staff.

Positive attitudes

A good positive attitude from the shore and ship employees makes a big difference to the way in which the tools of the SMS work. This attitude can only be achieved and continued if the commitment from the top is there.

If the senior most executives in the Company do not believe in what the system is trying to achieve then this will set barriers to the way in which the information flows are dealt with and will inevitably feed the doubts back to all levels of staff; however if the senior most executives do believe then this will also feedback to all the staff and will give more momentum to the processes of continual improvement.

We used people with the right experience and attitude to help construct the system to ensure the positive attitude was integral with the system. Simon Murray — Technical Director, Ian Couper — Consultant, myself — DPA and Marine Superintendent, Captain Arturo M. Tonson and the members of the crew of the *Tarquin Vale* for the initial input from ship staff.

It is human nature to associate with people of like mind and the persons responsible for the hiring of staff will influence the kinds of staff that will be employed and this must be borne in mind when selecting personnel managers.

It is essential not only to build the right team of people, but also to clearly define the lines of authority and responsibility within a company. If members of the team are unclear as to their or other people's role then this uncertainty will cause barriers to information flows and may lead to serious misunderstanding.

Information flows, reducing the barriers, transparency

The flows of information within a company are vital. They are the means in which control is maintained to allow people to do their jobs in an informed manner. They are also vital because all risks and hazards that are identified can then be dealt with, preferably before they become serious problems to the fleet.

Information flows are always at least two way flows. If information is only going one way then it is not working as effectively as it should be. Feedback on information is always required.

When determining methods of communication it should always be asked what are the key points of information that need to be exchanged between different parts of the Company to allow effective and efficient communications and control? The means of communication and its manner are important; any method decided upon must be reliable and of relevance to the importance of the information.

The right persons need the right information. Too little leads to ill informed decisions and too much to persons being overwhelmed with clutter, which in itself forms barriers to the effective flow of information. Concise and correct information should always be used that allows for easy understanding.

Information filters must also be considered. As with dealing with an emergency, the decision makers must have the correct information to control any situation but must not be overwhelmed with the other information that needs to flow when dealing with the specifics of a situation. For example, in a fire on board the vessel the master may instruct the head of the emergency team where a fire is, how he recommends the fire to be controlled and what route they should use to enter the dangerous area: he then does not need to know how the emergency team deal with the specifics but will need to be updated on their progress.

For each process within the Company it was clear that the decision makers for that process, and the lines of authority (the capacity for authorship — to write or rewrite processes), needed to be clearly defined. Other levels of monitoring were also to be established to help prevent the one man errors.

Any barriers to the information flows, such as doubt that any good will come from a report, or that it may not be read and acted upon, must be removed. For example, our reporting mechanism for accidents, incidents and dangerous occurrences (near miss reports to some) was an easily faxed or emailed page that was responded to by the Company, as soon as practicable, including looking at the root cause to any report and how it applied on not just the individual level but across the whole fleet. If necessary a fleet circular would then be issued to address any matter of safety arising and to try to prevent a further occurrence of the hazard.

With safety reports and cultures, the use of monetary rewards to conceal (or reward lack of reports), accidents, incidents and other reports, will always lead to information being restricted or distorted as staff try to protect what they see as their income. In TGC we made a commitment to maintaining a reasonable level of income with the salary of the staff and that the reporting of safety issues was an encouraged job requirement of all persons within the Company that was not linked to monetary rewards.

Another barrier that had to be removed was the culture of blame. This does not mean that there was no responsibility for actions but that all issues would be dealt with, without the heat of instant, ill-considered reaction. Most issues and problems on board result from training gaps; ignorance was our greatest enemy. Successfully applied risk management and uninhibited information flows protected us all from this ignorance.

One example of the information flow mechanisms working well within the Company occurred when one of our vessels was leaving port after having undergone main engine maintenance. The propulsion was controllable pitch and was fully tested as per the Company procedure before departure and found satisfactory. However once the pilotage was under way the main engine went to slow down with no apparent reason. The problem was quickly solved in the short term, as it was easy to reset the engine to maintain power. The superintendent on board talked with the master and it was decided that it would be safest if the vessel proceeded to anchorage whilst the root cause of the problem was diagnosed and cured. In the interests of safety the master talked with the pilot and they decided that it would be best to proceed with a tug to the anchorage rather than just on the ship's own power. The pilot called a tug and the ship was anchored.

Once the vessel was anchored the tug master produced an LOF form for the master to sign, which he, of course, refused to do. The master immediately called the office for advice and as the vessel had not been in immediate danger and the tug was employed on a risk control basis only, it was decided not to sign the form.

However, it soon became clear that the tug company was going to take the matter to court if necessary. The resultant settlement, after hefty consultation, meant we had to pay a quarter of a million dollars for the tug hire. The master was very anxious about the news and was wondering if it would affect his career. The technical director, realising that the incident could be misinterpreted, immediately called up the master to personally thank him for his professionalism in controlling the risk and to let him know that he made the right choice, whether or not the incident was then unduly complicated. This ensured that the Company safety management principles remained intact despite the cost and, I'm sure, would pay dividends to the Company in the years to come by furthering the trust and professionalism within the fleet.

The shortening of information flows to the decision makers is also necessary, the longer the route between one end of the communication chain and the other end (and vice versa) the greater the chance for failure or misunderstanding of the information and hazard control process. This does not mean that information of an important nature is not disseminated to all levels in the Company but that regard is always given to the effectiveness of those flows.

The role of the Designated Person Ashore (DPA) must be understood. The DPA should, where possible, be outside the routine running of the Company. He is there to safeguard the SMS processes and should be able to look at any aspects of the system to ensure they are continually being used effectively. In addition, all levels of staff, particularly ashore, should understand that this role is vital to effective management and provides a protection against the one man errors that can exist anywhere in the Company. The DPA must also have established paths for information flows to the uppermost parts in the Company to ensure that matters are understood by the top management and they can have the necessary information for effective management of the safety of the Company and its operations.

There is always a worry about company culpability with the information flows being unrestricted. Provided that legal bodies understand the principles of the ISM Code then I argue that full transparency with appropriate feedback is a better defence in an enquiry than to knowingly turn a blind eye to any hazards or risks.

Risk management is key

The principles of risk assessment, in their simplest terms, are to identify the hazards, determine the best way in which to manage the hazards, preferably removing them altogether, then to ensure full understanding by all persons involved. If these principles are applied to all processes in the SMS then it becomes easier to see the purposes of the processes.

> Risk Management = Protection from Ignorance

To better define what we considered as risk management, the definitions from the UK MCA's Code of Safe Working Practices for Merchant Seaman was used. Hazard is a source or situation of potential harm or damage. The following elements define and control the risk.

1   Chance that harm will occur.

2   Level of harm of such a hazardous event.

283

3   Level of risk.

4   Risk reduction methods.

5   Contingency plans.

Identifying hazards by safety and cost

To determine the particular areas that needed to be managed for our fleet it was possible to see from existing reports on safety and budget variances where the problems were.

By looking at the frequency of reports concerning accidents, incidents and dangerous occurrences from our previous system we could see that the pyramid required was not being achieved. Improvements in information flows and encouraging blame free reporting were needed.

Another example; on one of the vessels the main engine flywheel had abnormal wear. Luckily the problem was noticed before it became too serious and a major incident was prevented. After investigation it was found that the planned maintenance system being used did not give adequate protection to this part of the main engine and this was corrected for the fleet, preventing reoccurrence on not just the one ship but throughout the fleet. In fact the whole of the planned maintenance system was redone to allow for better control and for the possibility of continual improvement. A dedicated controller was appointed to maintain control and to help bring about improvements in the safety management system. This appointment was considered to be a vital element in maintaining the necessary information flows required by the Company for effective management of the fleet.

Another incidence was the cost of deviating for charts that should have been on board. We calculated that the cost of this was over a quarter of a million US dollars, including loss of hire. A full review of the procedures and a change of chart and publication suppliers, after the auditing of the new company to ensure good service, reduced the following years variance to only tens of thousands with a potential future improvement to be continuous.

How many of these out of budget variances occur within your company? Do you address root causes or just continue paying the bills? A good SMS is one that is self-correcting for these occurrences, preventing the reoccurrence to the best of known ability by dealing with the root causes. Figure 14.1 is a percentage graph of our change of variance, which can be attributed to an improvement of the information flows, and risk controls within the Company as provided by the SMS.

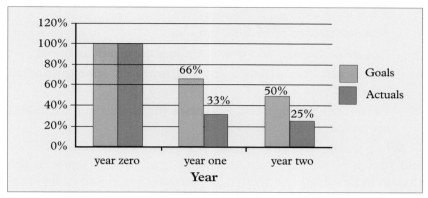

Figure 14.1 — Out of budget costs

Put your figures against the percentages to have an idea of how much it is possible for you to save with out of pocket costs (technical, insurance and commercial).

Once the problems were identified, then a course of corrective action could be determined to manage them. Moving from firefighting problems to initiating corrective actions to prevent reoccurrence is very satisfying and much more effective. As Granny said, "*Prevention is not only better but cheaper than cure*".

Determining the root causes requires the skill and professionalism of the Company staff and their understanding of the principles involved. If only a tenuous lip service is paid to the root cause determination then no effective solution can be found to a problem. If a procedure within the Company is being ignored or not properly understood then it should be looked at to determine why it is not being used not just to say it should be followed. It may be that the staff responsible for it's monitoring do not understand why it is written the way it is or that it is not relevant to them. Feedback from staff on the processes they use can highlight these weaknesses in a system before they become a significant problem.

Writing the SMS for the user

With top managements belief and commitment there has to follow all the staffs' belief and commitment and, when writing the SMS, thought should always be given to the following list:

1   Keep it simple.

2   Keep it effective.

3   Keep it relevant.

4   If its not needed then don't put it in.

5   Ensure authorities and responsibilities are clearly defined.

6   Ensure monitoring (double checking to prevent one man, human errors) is established.

7   Ensure risk management is in place (highlight the hazards and ensure information flows work).

When writing the SMS, remember whom it is being written for and why. It is not necessary to write an epic when a sentence will explain things. How it is written, the language used and its layout is also important. A cluttered script will not be conducive to quick learning and retaining the information. Colour, diagrams and the use of easily remembered phrases can enhance how the human brain absorbs and learns the information it is dealing with.

As an example of this we introduced Safety, Operations and Welfare (SOW) meetings instead of just safety meetings, because life at sea is a balance of all these aspects and they must always be considered together, there is too much compartmentalising at sea. All staff, whenever possible, were required to attend these meetings which allowed for interaction and training opportunities as well as full understanding for all staff on risk controls to manage recently identified hazards.

A greater understanding of the overall way in which a ship works is much more effective than to work independently and at cross purposes. Senior management both on shore and on board must always maintain the overview. In TGC we decided to produce two paper copies of the manuals for each vessel and a computerised version complete with hyperlinking to make finding the way around the system as easy as possible. The links on the CDROM version allowed for fast reference and when we linked the system to the correlation of the ISM Code, each aspect of an auditor's requirements were only a click away.

Effective tools and keeping it simple

Remember that the tools of the safety management system must be effective and simple. If a tool or process is complicated then it has more chance of going wrong or, even worse, of not being used properly.

If a tool or process is effective and credible it is more likely to be used. Seafarers are professionals and should be given the best support and tools to allow them to do an effective job.

Internal testing and review

As the SMS was being written we took time to consider how it all would work together and to test the mechanisms with ships' staff, taking their input and suggestions into account before the system was implemented. Feedback from the people, who were going to use the tools, was invaluable in assessing their effectiveness.

Authorised organization and approval

Once the system was constructed and tested then we moved to the next step of obtaining approval from the Flag States authorized organization, for us Bureau Veritas classification society document review, followed by interim auditing. We wanted an organisation that would be fully aware of the principles of the ISM Code and have the resources to provide knowledgeable auditors throughout the world to cover our ships. If we were going off on the wrong line we needed to be told so, that we could correct our procedures and ideas.

Keeping ahead of the game, legislation monitoring, research and development

The SMS must conform to the ISM Code and stay ahead of current and upcoming legislation. The means to stay ahead of any upcoming changes need to be determined and regularly revisited. New techniques and ideas are always being suggested within the industry, some are very good, others not as much. If a new system or process comes along that suits the Company and can make the systems and processes easier and more effective, then it needs to be researched and, if approved, can be incorporated. Research and development is the lifeblood of any company in any industry.

Emergency response and creating an effective team — training

A large part of any SMS is the system for ensuring the Company can handle emergency response situations. As with all of the SMS we put together, we stuck to the principle of keeping it as simple and as effective as possible. Risk assessment again formed the backbone of the emergency response system. Information flows needed to be reliable, flexible and fast. Identifying the hazards and ensuring all parties knew of the risk controls necessary were the keys.

From lessons learnt from previous accidents it was known that to deal effectively with any emergency, the decision maker, the person appointed for controlling any situation, must be the one best suited to keep the overview of the situation and must have the authority to effectively deal with the situation. We had several individuals who could assume the role and a strong pool of knowledge of all the aspects of the industry from which we could gather the information and ensure effective and relevant control of any situation.

In support of the decision maker a good team of advisors was needed. The basic principle would be to relieve as much of the stress and workload from the master and crew as possible to allow for better control of the situation at the scene of the emergency. It must be remembered that the ship's staff are at the sharp end of an emergency and would be under a lot of stress that interferes with their decision making abilities. A team of possible experts was also set up that could be flown out to the scene.

A system was set up to simplify the means of initiating an emergency response with only one number to call which was forwarded to whoever was on duty in the Company. The flow diagram at figure 14.2 explains how our system would work in an emergency.

We had one contact number for anyone to call that was manned at all times and would simplify the ship's staff contact needs in an emergency or doubt about an emergency. The duty person was always assigned and a notice in the office detailing who was assigned was posted at the entrance to the technical department.

Once the system was established the testing of the systems and the training of the use of the information flows and decision making processes needed to be ensured with regularly drills that tested specific areas where training gaps existed or processes and resources needed improvement.

Implementation planning

It was very clear to us that the best way to ensure that the SMS was not only used but also understood and encouraged was to introduce the system personally to all members of our staff. The training of shore staff, and of the ships staff (certainly all senior officers), in the

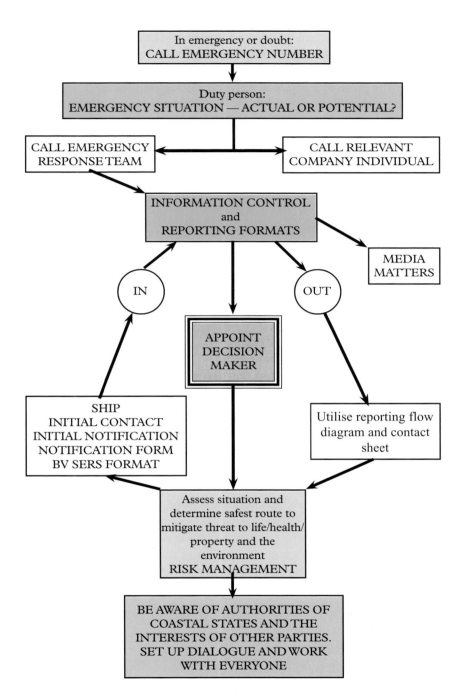

Figure 14.2 — Emergency response system

system, was conducted, by us, on a one to one basis. This gave us the opportunity to ensure that company employees saw how they played a part in the SMS and why the tools were there. It also allowed us to nurture the positive attitude at all levels throughout the fleet to ensure the information flows would work. To achieve this we sailed on each vessel prior to taking over the technical management and conducted the training on board with all ship's staff as required. In addition the training master was able to conduct training in Manila for new joiners prior to embarkation.

How, why, proof

A fundamental principle of the auditing process and for training is simply to explain to the staff how, why and the required proof (which allows for data to be collected and systems and processes to be reviewed). With full understanding the crew were empowered with the correct knowledge and attitude to use the tools of the SMS.

Risk assessment and information flows

A major part of the implementation was the explanation of the new ideas and how we wanted our SMS to work. The crews needed to know about risk assessment and the need for unrestricted information flows. They needed to know that the shore staff were there to give support when needed and to help plug any training gaps found in both individuals and the fleet as a whole.

Maintaining the overview

With all processes, and for keeping a good system for monitoring safety, maintaining the overview is vital. Those in positions of authority should be given the correct information for doing their jobs and be allowed the time and space to maintain the overview of what is happening so that they can spot problems before they become serious. If this overview is lost, either on board or ashore, then problems will not only arise but they may become very serious.

Training the trainers

One area of our work that is not fully taught during the statutory training, particularly at sea, is the training and equipping of the management team on board the ship on how to be effective managers and trainers. There is little enough time during voyages in the modern merchant navy that has been cut back to the bear minimum in an effort to save costs and remain competitive, without supplying the staff we do employ with effective training in all the aspects needed, particularly in light of the requirements of a good SMS. This deficit in statutory training is an oversight the industry needs to correct.

This just briefly leads me on to one point on risk management that seems to have escaped a lot of the industry: If we are trying to prevent accidents and incidents, to make things safer and to protect the environment from pollution we need to know the hazards. The greatest hazard is known to be human error, particularly one man errors, yet the industry has gone down the line of reducing crews and so increasing the frequency of exposure to one man errors.

Can this be right? It is certainly against the principles of good risk management, because if you cannot eliminate the risk, you reduce your exposure to it and control the situation effectively and with full knowledge. Does this not make sense to you? Yes? Good, since this principle concerning the hierarchy of risk management is the cornerstone of health and safety law in the UK and many other countries. What is the cost of an extra seafarer when compared to the cost of a total constructive loss with the associated knock on effects?

To enhance training on board and to ensure that specific training gaps were plugged we decided to introduce a system of rolling trainers from the fleet. Captains and chief engineers who would be given enhanced training in the areas needed and then once fully trained would be required to go around the fleet visiting each vessel at least once a year to ensure that training was conducted on board during the voyage, when time gave us greater safety and a better learning environment.

The additional training, which the training masters and chief engineers were given, was adapted to the particular requirements and issues at the time of their service, but included regular visits to the head office in Edinburgh so that the culture and attitudes on shore could be fed back more effectively to the vessels.

The bonus of this rolling system was that over the years the enhanced training would lift the general knowledge base in the Company and improve our protection against ignorance and misunderstanding. In a relatively short time we would have many masters and chief engineers with this enhanced training throughout the Company. That is not to say that many of our seafarers were hesitant to assume this role until they understood it more clearly.

Setting the culture

The trust and respect of the crews was considered to be very important. We had crews from the Philippines on all of our vessels, some of whom had been with the owner for over nineteen years; others were new to the Company. Nurturing the proper levels of feeling and understanding within the Company was vital to our success. We needed crews who knew and understood our system and our attitude and would make a meaningful contribution to the SMS. In return we provided good conditions, a sense of belonging, and the knowledge that we, as technical managers, were listening and responding to our crews. In short we created a sense of family. I believe we were largely successful in this, as testified by the constant positive feedback and improvements we were able to make in the Company. I would like to thank the crews of Tarquin for all of their support and hard work: It was a great privilege to work with you all. Maintaining a positive attitude on the vessels and ashore is needed and if some useful information is given then a polite thank you with the feedback will always go further in setting the right culture than an unnecessary shouting match when things go wrong.

During the implementation of the system we identified one particular master who, from his previous correspondence, might have proved difficult to bring to the right attitude. The implementation officer went to this vessel and found that during the course of the training the master's initial inert and negative attitude suddenly changed as he saw for himself how the system would work and very quickly he was very enthusiastic about the system. This ability for the system to sell itself to those using it was a reward to us and increased the enthusiasm in the office. It would make running the system that much simpler, but was only a starting point. Once this base culture was set, it needed to be maintained. The continuity of the crews is very important, if you make the investment into training of the personnel you want to keep them for the long term.

The auditing process

Auditing processes are very necessary. The verification of the effectiveness of any system and its continued improvement provides valuable information to the management team. For the internal auditing we needed flexibility and consistency. We externally trained all of our shore staff as auditors and this provided a fundamentally sound base of knowledge on the ISM Code. We then trained all of our ship visiting staff on the principles of how to train others.

Our first rounds of audits were mainly done in port but we found this to be limiting in the needed interaction to properly assess the system. Our plan then became to conduct internal auditing at sea, during one of the sailing visits we had included already into our system (at least four a year by different members of the management team). This provided the opportunity not only to carry out the internal audit in a safe and practical manner but also allowed for immediate follow up action by the auditor to plug training gaps that came to light.

In addition to auditing ourselves we instigated auditing of service providers; crew managers, suppliers of charts, IT support and other aspects upon which the Company needed information and to confirm their reliability. This additional auditing could only be done because of the number of auditors available to us at any given time.

The SMS familiarisation booklet

To further enhance understanding and appreciation of what we were trying to achieve we produced an SMS familiarization booklet in which we explained briefly the principles of our system and how it would work. This booklet was given to all seafarers in Manila prior to an assignment and was used during on board training sessions. Below is a quote from the front of the booklet.

"Principles

It has become evident that a large proportion of accidents and incidents that occur at sea happen because of failures in good practice due to human error. These SMS procedures have been put in place to safeguard our operations against such occurrences. Basic good common sense, professional knowledge and the need for double verification in all aspects, together with positive attitude and effective communications between all parties are our keys to safe operation.

Information Flows

Information flows between all parts of the Company are essential for the effective running of a safe and efficient system.

- This means that all appropriate persons need to be informed of all relevant information to enable them to be more effective in their jobs.
- The various meetings on both ship and within the Company provide a mechanism for this information exchange and all managers are expected to use these tools for the improvement of all.
- The communications from ship to shore and from shore to ship are to be used to pass on any relevant information such as defects

and problems. In this way any problems and defects can be addressed and rectified as early as possible.

If you are already on board the ship when it is taken over by Tarquin, you will have discussions directly with the Company implementation officer for your ship using a one to one format. The objective is that you fully understand the system and how it should make your life safer. It may be that you feel that the company's programme of SMS does not fully cover all aspects, or that there are aspects, which are perhaps incorrect."

> YOUR VIEW IS VERY IMPORTANT. THAT IS THE HEART OF ISM — THAT YOU, WHATEVER YOUR POSITION ON BOARD, ARE FULLY INVOLVED AND THAT YOUR OPINION IS HEARD.

Continual improvement; a living system

To show how the system fits together for inputting information for continual improvement we constructed wheel diagrams (figures 14.3 and 14.4 — see overleaf) for the SMS and for training. The wheels give the mental image of movement and progress.

Key indicators

Key indicators are well known in the industry. They are there to help measure systems and their effectiveness. Some key indicators for the SMS are the measures of the information flows. It could be argued that the single most important are the reports for accidents, incidents, dangerous occurrences and non-conformances. The pyramid of these reports gives a very quick visual measure of the effectiveness of the SMS. Shown in figure 14.5 is our pyramid for last year of operation.

To ensure that non-conformances were dealt with in a timely manner a defects, observations and non-conformances (DNCO) list was started for each vessel. This would include any item from an external inspection, any major item or repair or interest noted on board or ashore that required shore interaction. The DNCO list was in simple terms our management focus for both the ships and the shore staff responsible for each vessel. It was not the only, nor even the primary, means of reporting any problems but gave us the back up to prevent missing any instances from one man errors.

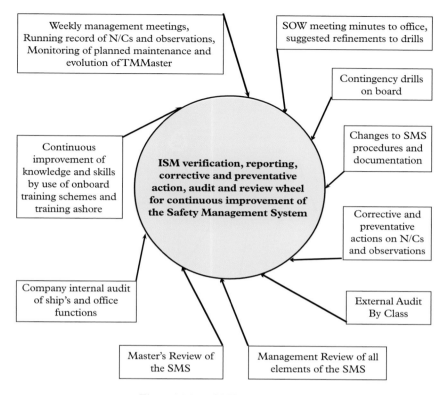

Figure 14.3 — SMS wheel diagram

From the list we could see if problems existed on more than one ship and the list gave us another effective tool for managing the vessels and for spotting any areas of risk, within the fleet, that could lead to problems.

It was also a method of information flow and a column for office feedback was introduced after suggestion by ships' staff to ensure the information flows worked two ways more effectively.

There is a danger sometimes with key indicators and setting the required levels. How many dangerous occurrence reports, for example, should a ship send in every year? Does more mean that things are not good? Who decides what are the correct levels? Is there a political, commercial or a legal reason why the number is set where it is by management? All these things can be barriers to free flowing information. The important thing for us was not to go in with preset ideas and set targets but to encourage all reporting and to use the information gathered to correct any problems. That will automatically lead to greater

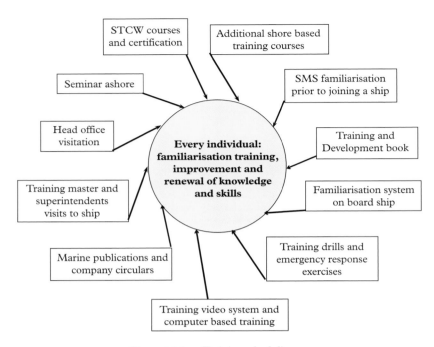

Figure 14.4 — Training wheel diagram

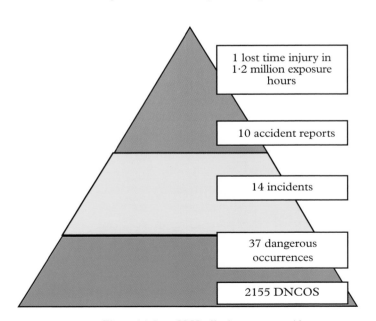

Figure 14.5 — SMS effectiveness pyramid

296

safety by the quickest and most effective means. Setting limiting key indicators levels will add to information flow barriers and give only an apparent aura of safety but will not give the true level of safety.

Review; effective measuring

One, of many important lessons, that was taught to me by Phil Robson from Bureau Veritas was how we should look at reviewing. There are many reviews carried out in technical management of ships and with the SMS; management reviews, masters' reviews, training drill reviews. All of these tools are needed for improvement but how do we make them effective? The key was in understanding that a review is an effective measure. If what we are writing in our review is not measuring the subject matter effectively, whether for good or bad, then it is not a good review. With any review, regard should always be given to ensuring adequate resources; personnel and or equipment to increase the simplicity or effectiveness of a process; or adequate review of the processes to ensure that they are effective with the available resources.

Spotting the training gaps

Training gaps are there in any industry and are the misunderstandings and ignorance that are the enemies of the safety culture. We needed mechanisms and awareness to spot them as early as possible in order to ensure the proper information was fed back to the crews, or the correct training was implemented to the right people to protect the fleet from such hazards. We used all reporting mechanisms including personnel assessment forms to highlight any weaknesses. These were then considered and discussed and possibly resulted in either fleet circulars, Tarquin training sessions, shore team visits in response to particular issues on board, or training (or retraining) ashore.

Seminars and shore based training

In addition to the implementation training and the on board training from the training master we had a seminar in the Philippines about once a year and on the last occasion I was joined by 23 of our senior officers who came to Manila for a few days, hosted by our crew managers Pacific Ocean Manning. The itinerary of our interactive sessions together is overleaf (figure 14.6) and you can see how we built upon the earlier training. The positive heart and sense of team purpose at the close of the seminar was a reward in itself.

During the course of the seminar I issued a multiple choice questionnaire which had been constructed by the shore team with particular attention to areas of the SMS that had been noticed as being in need of improvement or greater understanding. The idea was to have a measure of the actual knowledge of our seafarers and how well we had done in training them. It would also give me another opportunity to re-emphasise some of the weaker aspects and areas as we had found them from the information flows between ship and shore. I was pleasantly surprised to find that the average mark for the questions was over 75% and, when we went through the answers, the interaction between all of the people in the room was lively and very positive.

To ensure that the training given by shore colleges and training centres was of the correct standard we arranged to visit and interact with these training centres. This interaction was very beneficial for understanding how training succeeded or if it had any weaknesses that could be corrected. For example we noticed a weakness with the master/pilot information exchange and interaction, after talking with the training centres in the Philippines they responded by adding greater awareness of the master-pilot relationships to their training courses for Tarquin Employees. Another weakness was noted in understanding the need for positive reporting particularly during stressful situations. Positive reporting means having the correct information flows to control a situation or action. The easiest and most commonly understood example is that of the helm orders.

When the master gives a helm order, the order is repeated back to the master to ensure that the order is understood. The action is then carried out and once finished the information is given back to the master so that he knows the action has been completed.

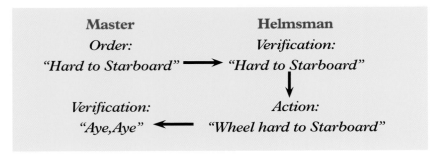

If this method of order and action, with verification, is carried into the emergency situations then the master can have effective flows of information and be able to make the informed decisions about any situation.

Tarquin Senior Officers Seminar — Manila 2002		
Captain Sean Noonan **Marine Superintendent and DPA** Captain Arturo Tonson Fleet Training Master		
Times	**Date: 20th** Wednesday	**Date: 21st** Thursday
0900—1000	**Opening Introduction:** Why and How of ISM? {SN}	**Oil Record Books** {SN}
1000—1030	*Coffee*	*Coffee*
1030—1130	**Training the trainer and training drills on board** {AT}	**Efficiency of operations** {AT}
1130—1230	**ISM - is it working?**	**Informal open discussion** {SN/AT}
1230—1330	*Lunch time*	*Tarquin Lunch*
1330—1430	**ISM Workshop** {SN/AT}	
1430—1500	*Coffee*	
1500—1600	**Tarquin Accident and Incident Analysis** {SN}	**Media Training**
1600—1700	**Risk Assessment and Safety Culture** {AT}	
1700—1800	**Company SMS multiple choice**	**Closing Meeting**

Figure 14.6 — Tarquin Senior Officers Seminar — Manila 2002

299

Ship based training

As previously mentioned, we had taken training and the plugging of training gaps in the fleet very seriously. All of the shore team were trained to be or were about to be trained as trainers. In addition, the training master and training chief engineer were also given training in how to conduct effective training. A set of additional training sessions to the usual Videotel training videos and company training as provided by the crew managers was set up called Tarquin Training Sessions. These were designed for use at sea by any superintendent and were specific to our company's training needs.

Ensuring information flows work

To ensure that the information flows are working effectively the following flow diagrams show in simple form how they are meant to work in respect of risk management.

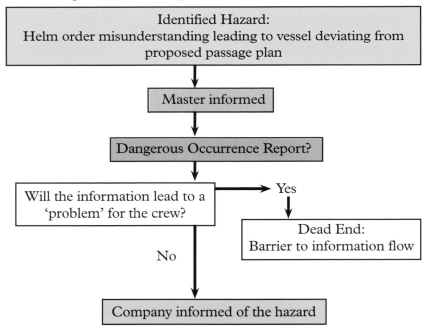

Figure 14.7 — Information flows in risk management

In this instance you can see that the identified hazard leads to information flowing in one direction. However to have effective information flows we need the information to work in more than one direction. Feedback is vital for effective management.

In the following diagram we can see how the information flows work if we feedback correctly.

From this diagram it can be seen, that if set up correctly, information flows lead to automatic self-correction, prevention and, therefore, effective risk management.

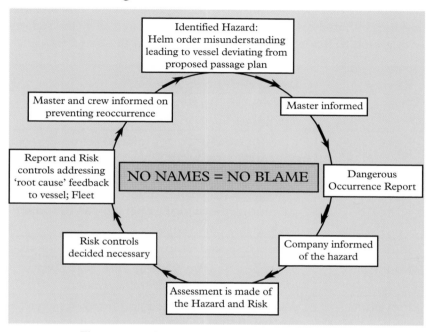

Figure 14.8 — Information flows if feedback correctly used

Dangerous trends

From the information flows we received, particularly the DNCO lists and accident, incident and dangerous occurrence (AID) reporting we were able to spot dangerous trends of misunderstanding in our fleet and, by quickly responding to the up to date information coming from the ships, we were able to improve everyone's awareness and so effectively manage the perceived hazards.

One such trend noticed was the failure of junior officers to properly oversee the safety of mooring operations or lifting operations. This was found to be a training gap in the preparation of the junior officers for their role on board. An immediate fleet circular was issued, once the full extent of this training gap was evident. This was followed up by the compilation of a Tarquin Training Session on mooring operations and lifting operations with particular emphasis on the role of the safety man.

Once this information was sent out to the fleet, the dangerous occurrences, incidents and accidents were greatly reduced in respect to this issue (i.e. our exposure to the risk was reduced). Verification of the effectiveness of this preventative action and its full understanding was to be obtained during company internal auditing.

To ensure that these lessons were not forgotten, a quarterly accident and incident analysis was produced which revisited all of the reports of that month. Below is an example from one of these reports.

Reviews	Dangerous occurrence
Details	One mooring winch tripped off during mooring operations
Risks	Loss of resource for effective mooring operations causing a potentially unsafe condition for the vessel and crew
Cause	Emergency stop switch unintentionally banged into during the mooring operation on the poop deck.
Initial actions	System reset and returned to normal operations
Root Cause	Emergency stop switch not sufficiently protected against accidental activation.
Prevention	A guard to be fabricated and fitted to protect the switch (that will still allow for quick access) and the switch is to be better highlighted to help make people more aware of its location.

Figure 14.9 — Quarterly accident and incident analysis

Uncontrolled costs

On the technical side of the operation we regularly reviewed the budgets and the variances to see if any aspects of control had been missed. The better control and root cause solving of the problems would lead to more effective use of the available money to run the vessels.

Follow up and verification

With all systems that are set up to continually improve we need to know that improvements are being made. Not all change is for the better. Keeping things simple and easy to use is fundamental.

Sometimes some of our processes were found to be written in too complicated a way, which led to misunderstanding. We always kept an eye on how things were working and how we could improve the tools of the SMS.

Only by following up on the initial training and by verifying the data from reports and reviews, whether internal or external, could we see if the mechanisms for continual improvement were working.

Every six months we conducted a full management review; an effective measure of systems and whether they were working. During these reviews we looked not at making ourselves feel good but at trying to objectively look at all the key processes for the SMS and how they were working. Also by using the feedback from the ships we took into account the ship's staff perspective.

Throughout the running of the system we still encountered inertia against free and open reporting, the legacy that is inherent in the industry for hiding problems out of fear, either of looking like a fool, or of keeping the job, is a very hard one to overcome. To keep improving the information flows and to ensure that the right culture of thinking was always re-emphasised the visits from the shore staff and the senior officers' visits to the office were of high priority: The sailing visits by us particularly helped cement the bonds and the trust.

Change control

The control of any changes to the SMS was a great area of risk for us. After having gone through the trouble of ensuring understanding and compliance to the original system we needed to maintain the high level of understanding within the fleet with any changes that were made. This was done by several means.

1   Making the changes easy to find and identify.

2   Making the changes concise and easily understandable.

3   Ensuring the processes of testing and approval for change were clearly defined

4   Ensuring that whenever possible senior officers were routed through the head office to undergo updating training.

5   Ensuring the continuation of the on board training during the sailing visits by the shore team and by the training master/chief engineer.

Finally, for us, the keys to success were to ensure that the culture and attitudes were correct at all levels of the Company. To ensure that risk assessment was always included in all processes and that the information flows worked to and from all parts of the Company.

We did not get everything right all the time but by being open we could see where the problems lay and could take action to correct them at root cause level. I feel we had a great system but it was only in its infancy, any system must be capable of improvement.

For anyone who undertakes to run a SMS you must believe in it to make it work. It can be done. We in Tarquin made considerable improvements in attitude, safety and cost by treating the system as the mechanism that ran the whole operation.

Thoughts for the future

Although we were a long way from perfect we did manage to achieve significant improvements by changing the culture of thinking on board. Our actions were mostly reactionary however, leading to corrective response to circumstances that had already occurred. A greater level of safety is possible with more preventive thinking, some of which we had with the use of our risk assessments.

One area, which needed more understanding and encouragement, was in the spotting of the warning signs that are always present prior to an accident's occurrence. If the Company as a whole can focus on identifying these warnings and by use of timely information flows and risk controls can prevent them occurring, by using all persons, at all

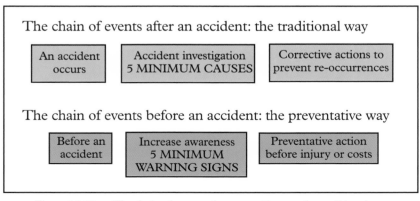

Figure 14.10 — The chain of events after an accident — the traditional way

levels in the Company then this would be the best way to achieve a collective preventative safety culture.

Prevention is better and cheaper than cure.

In pictorial form this can be explained as in figure 14.10

In your system, is there room for improvement, with these thoughts and ideas in mind?

Good luck to you all and bon voyage.

BIBLIOGRAPHY

Primary Source Documents

IMO Conventions and Resolutions

- International Convention for the Safety of Life at Sea, 1974, as amended Chapter IX – Management for the safe operation of ships
- Resolution A.443(XI) Decisions of the shipmaster with regard to maritime safety and marine environment protection
- Resolution A.741(18) as amended by MSC.104(73) International Management Code for the Safe Operation of Ships and for Pollution Prevention (International Safety Management (ISM) Code)
- Resolution A.913(22) Revised Guidelines on implementation of the ISM Code by Administrations (Revokes resolution A.788(19)
- Resolution A.848(20) Implementation of the International Safety Management (ISM) 1997
- Resolution A. 852(20) Guidelines for a structure of an integrated system of contingency planning for shipboard Emergencies
- Resolution A.880(21) Implementation of the International Safety Management (ISM) Code by 1 July 2002

IMO Circulars

FSI

- FSI/CIRC.9 927.11.2000) List of non-governmental organisations authorised to carry out surveys and issue certificates on behalf of Administrations

MSC

- MSC/Circ.693 (26 May 1995) Draft amendments to the International Safety Management Code
- MSC/Circ.761 (11.7.96) Timely and effective implementation of the ISM Code
- MSC/Circ.762 & MEPC/Circ.312 (11.7.96) Guidance to companies operating multi-flagged fleets and supplementary guidelines to administrations.
- MSC/Circ.771 (12.12.96) Implementation of the International Safety Management (ISM) Code

- MSC/Circ.828 & MEPC/Circ.334 (7.11.97) Implementation of the ISM Code and interim documentation
- MSC/Circ.881 (14.12.98) Implementation of the International Safety Management (ISM) Code by 1 July 2002.
- MSC/Circ.889 & MEPC Circ.353 (17.12.1998) Self – Assessment of Flag State Performance
- MSC/Circ.890 Interim guidelines for port state control related to the ISM Code
- MSC/Circ.927 (21.07.99) Documents of compliance issued under the provisions of the International Safety Management (ISM) Code. 1999
- MSC/Circ.954 & MEPC/Circ.373 (23.6.2000) Self-Assessment of Flag State Performance : Criteria and Performance Indicators.
- MSC/Circ.994 MEPC/CIRC.381 (1.5.2001) The beneficial impact of the ISM Code and its role as an indicator of safe operation and environmental protection.
- MSC/Circ.1015 (12.6.2001) Reporting Near Misses
- MSC/Circ.1020 / MEPC.Circ.387 (21.1.2002) Implementation of the International Safety Management (ISM) Code by 1st July 2002

UK Statutory Instruments
- 1998 No. 1561 – The Merchant Shipping (International Safety Management (ISM) Code) Regulations 1998
- 1998 No 1838 – The Merchant Shipping (Code of Safe Working Practices for Merchant Seamen) Regulations 1998
- 1999 No. 2567 - The Merchant Shipping (Accident Reporting and Investigation) Regulations 1999

Marine Guidance Note
- MGN 115 (M+ F) MAIB – Accident Reporting and Investigation

Reports and Inquiries
- BIMCO / ISF 2000 Manpower Update – The Worldwide Demand for and Supply of Seafarers: Main Report – April 2000 – Institute for Employment Research (IER) – University of Warwick

- Department of Transport – Marine Directorate – The Human Element in Shipping Casualties – HMSO - 1991
- Department of the environment, transport and the regions – MAIB – Memorandum on the Investigation of Marine Accidents - 2000
- House of Lords – Select Committee on Science and Technology – Safety Aspects of Ship Design and Technology. London. HMSO. 1992
- UK P&I Club – Analysis of Major Claims - 1991

Secondary Source Documents
IMO Website Press Briefings
- '78% of fleet set to meet target, says IMO' Briefing 9 April 1998
- 'ISM Code becomes mandatory' Briefing 1 July 1998
- 'ISM Code must not become a 'paper exercise' warns Secretary-General' Briefing 20 February 2001
- 'IMO welcomes ISM Code study: ISM Code – phase 2 ships' Briefing 25.9.2001
- 'IMO issues ISM Code warning' Briefing21.1.2002
- 'Shipping enters the ISM Code era with second phase of implementation' Briefing 28.6.2002
- '2002 marks 25th annual World Maritime Day – IMO – Safer shipping demands a safety culture' Briefing 26 September 2002

Commentaries / Guidelines / Miscellaneous Reports
- Aon – An Insurance Market Overview – May 15, 2002
- Chamber of Shipping of America – Environmental Criminal Liability in the United States – A handbook for the marine industry – September 2000 – ISBN 0 9702752 0 X
- The Code of Safe Working Practices for Merchant Seamen – Maritime and Coastguard Agency – 1998. ISBN 011 5523693
- Health and Safety Executive – Reducing error and influencing behaviour – HSG48
- Informa Prossional – P&I International – International Directory of Ship Arrest Lawyers 2001 edition – ISBN 1 85978 308 2
- International Association of Classification Societies (IACS)

- Guidance for IACS Auditors to the ISM Code. London : IACS 1996

- Interpretation of the International Management Code for the Safe Operation of Ships and for Pollution Prevention (ISM) Code, adopted by IMO Resolution A.741(18). London : IACS, 1995

- Procedural guidelines for ISM Code Certification. 3rd ed.. London : IACS, 1995

- Guidance for IACS Auditors to the ISM Code. London, IACS, 1995

- Synopsis of the IACS Model Course for Training ISM Code Auditors. London, IACS, 1995

• International Chamber of Shipping (ICS) – Assessment and Development of Safety Management Systems. Part I : Assessment Checklist; Part II : Example Documentation. London, ICS, 2 Vols. 1997

• International Chamber of Shipping (ICS) / International Shipping Federation (ISF) – Guidelines on the application of the IMO International Safety Management Code. 3rd ed.. London : ICS/ISF, 1996

• International Chamber of Shipping – Bridge Procedures Guide – Third Edition - 1998

• International Shipping Federation – Guidelines on Good Employment Practice - 2001

• INTERTANKO – Systematic approach to Tanker Accident Analysis – Lessons Learnt

• ITF – Globalisation – the cost to the seafarer – Commemorating World Maritime Day – 27 September 2001

• The ISF Year 2001 – The annual report of the International Shipping Federation

• The Lancet – Vol 360 Number 9332 – All at Sea – 17 August 2002

• Lloyd's Register – ISM for the Master-Owner small companies

• MAIB Annual Report 2001 – Chief Inspectors Foreword by Rear Admiral John Lang

• The Paris MOU – Port State Control Instructions – PSCC34/ 2001/01 dated 11 May 2001 Guidance for the Port State Control Officer on the ISM Code

- SIRC – Seafarers International Research Centre – Cardiff University – Transnational Seafarer Communities – ISBN 1 900174 17 0
- The Swedish Club Highlights – ISM's beneficial impact – December 2001
- Wharton School, Centre for Risk Management and Decision Processes - Near-Miss System Analysis: Phase 1 – December 2000

Other Publications
- Anderson, P. – ISM Code: A practical guide to the legal and insurance implications. London, Lloyd's of London Press, 1998. (ISBN 1-85978-621-9)
- Anderson, P. and Kidman, P – A Seafarers Guide to ISM. Newcastle upon Tyne, North of England P&I Assn. 2002 (ISBN 0-9542012-2-1)
- Anderson, P. and Kidman, P – What have the World Cup and ISM got in common? Newcastle upon Tyne, North of England P&I Assn. 2002 (ISBN 09542012-1-3)
- Botterill, G.J – The ISM Code – an independent view - Paper presented at the IMAREST Conference – ISM Code – making it really work'. London 13 / 14 May 2002
- Cuneo, B. – Practical implementation of SMS onboard - Paper presented at the IMAREST Conference – ISM Code – making it really work'. London 13 / 14 May 2002
- Donaldson, J Lord. – 'The ISM Code: The Road to Discovery?' Paper presented at the Inaugural Memorial Lecture to Professor Cadwallader, London 26 March 1998, The London Shipping Law Centre
- Goebel, E (Fleet Safety Officer with A.P. Moller) – A Shipowners perspective on the ISM Code – Paper presented at the BIMCO Residential Course in Copenhagen 3-5 September 2001 – The ISM Code and STCW 95
- Hernqvist, M. – The Swedish Club - A Hull and P&I underwriter's experience of ISM's beneficial impact - Paper presented at the IMAREST Conference – ISM Code – making it really work'. London 13 / 14 May 2002
- Juhl, R – (Safety and Quality Manager in Tschudi & Eitzen Ship Management (Denmark0 A/S) – A Ship Managers

perspective on the ISM Code - Paper presented at the BIMCO Residential Course in Copenhagen 3-5 September 2001 – The ISM Code and STCW 95

- Koch, E (Managing partner – Oesterreichischer Lloyd Ship Management) – ISM Code – Expectations versus reality in the development of an effective SMS – Paper presented at Nautischer Verein zu Bremen 8th seminar – ISM Code – Half time or additional playing time? – Bremen 24 February 2001

- Kuo, C. Managing Ship Safety. London. Lloyd's of London Press. 1998 (ISBN 1-85978-841-6)

- Langkabel, J – (Dipl Ing – Maritime Service Centre Manager – DNV Germany GmbH Hamburg) – How can a Safety Management System be developed and effectively implemented? - Paper presented at Nautischer Verein zu Bremen 8th seminar – ISM Code – Half time or additional playing time? – Bremen 24 February 2001

- Mandaraka-Sheppard, A – (The London Shipping Law Centre) – Reflecting on the ISMC: do judicial or legislative processes reinforce the original objectives of the Code? - Paper presented at the IMAREST Conference – ISM Code – making it really work' . London 13 / 14 May 2002

- Morris, P Hon. – Chairman – ICONS – International Commission on Shipping – Ships, Slaves and Competition. Charlestown NSW Australia. 2000. (ISBN 0-646-41192-6)

- Orrell, B – NUMAST - The ISM Code – paper or reality? – Paper presented at the IMAREST Conference – ISM Code – making it really work' . London 13 / 14 May 2002

- Reason, J – Human Factors, Cambridge University Press, 1994

- Rodrigues, A.J & Hubbard, M.C – The International Safety Management (ISM) Code: A new level of uniformity - 1999

- Sagen, A. – The ISM Code in practice. Tano Aschehoug, Norway, 1999. (ISBN 82-518-3825-8)

- Sekimizu, k – IMO representative - Keynote Address at the IMAREST Conference – ISM Code – making it really work' . London 13 / 14 May 2002

- Shuker, M – (representing the International Chamber of Shipping) – What the future holds for the ISM Code - Paper presented at the IMAREST Conference – ISM Code – making it really work' . London 13 / 14 May 2002

- Tatham, S. and Honan, W.J. and Faerden, A.W. – 'Legal Implications of the ISM Code' – Oslo – Intertanko

- Wadmark, O – (Senior ship surveyor – Swedish Maritime Administration) - Paper presented at the BIMCO Residential Course in Copenhagen 3-5 September 2001 – The ISM Code and STCW 95

- Witherington, S – MAIB - ISM – What has been learned from marine accident investigation? - Paper presented at the IMAREST Conference – ISM Code – making it really work'. London 13 / 14 May 2002

ANNEX 1

GOAL SETTING LEGISLATION IN SUPPORT OF SAFETY MANAGEMENT

Introduction

Historically, health and safety legislation has been prescriptive, i.e. setting out in detail what you must or must not do in order to comply with the law. Such legislation is often welcomed, both by those subject to it and some of the regulators responsible for enforcing it, on the basis that it is clear to everyone what is required. The weaknesses of such an approach are that it cannot cover every eventuality and every set of circumstances. It also tends to play down the important role of management and does not readily respond to or encourage technological advances.

In the UK, since the passage of the Health and Safety at Work Act 1974, a mainly goal setting approach has been adopted with respect to new health and safety legislation. This involves specifying a goal to be achieved, often qualified by such phrases as 'so far as is reasonably practicable' or 'so far as practicable'. The first involves balancing the risk to be managed against time, trouble, technical difficulty and cost involved whereas the latter does not take into account cost and is a much more stringent standard.

Such an approach allows those with duties under such legislation to adopt an approach of their choosing, provided that it meets the criteria of 'so far as is reasonably practicable' or 'so far as practicable' or similar qualification. This approach requires both the duty holder and the regulator to really think about the management of the risks involved in the management of the undertaking. However, for such an approach to work well, it needs to be underpinned by guidance which represents best practice.

In the UK, such health and safety legislation is underpinned by Approved Codes of Practice (ACoPS), guidance from the UK Health & Safety Commission/Executive, HSC Industry Advisory Committee and trade associations which provide extensive and valuable guidance as to what represents best practice. On the other hand, it does not prevent a duty holder adopting a different course of action so long as that course of action achieves a similar standard to that represented by the ACoP. The UK Merchant Shipping and Fishing Vessels (Health and Safety at Work) Regulations 1997 as amended in 2001 are an example of a largely goal setting piece of health and safety legislation in the maritime world.

The ISM Code is goal setting. It allows management to decide how best to achieve the objectives set out within it. However, for such an approach to be effective, it will be necessary to identify and promulgate best practice for the various sectors within our industry in the form of codes of practice and guidance. This is a marvellous opportunity for the whole maritime industry to come together and examine itself from top to bottom and use the well tried and tested goal setting route and move on from the prescriptive approach to which this industry has been subject for so long.

Goal setting legislation

Goal setting legislation sets out what must be achieved but not how it must be done. However, sometimes it is necessary to be prescriptive where what should be done is spelt out in detail.

The degree of risk in a particular job or work place needs to be balanced against the time, trouble, cost and physical difficulty of taking measures to avoid or reduce the risk. Essentially, the law requires what good management and common sense would lead employers to do anyway, which is to identify the hazards, assess the risks associated with those hazards and take sensible measures to avoid or control those risks. Modern health and safety law in the UK and much of Europe is now based on the principle of risk assessment.

Although a goal setting approach provides a good degree of flexibility in the way in which an employer has to achieve the defined goals, there is a need to provide appropriate guidance to duty holders. In the UK, the Health and Safety Commission and Health and Safety Executive, the bodies responsible for ensuring compliance with the Health and Safety at Work etc. Act 1974, use the options of guidance, Approved Codes of Practice and regulations and try to use whichever option allows employers most flexibility and costs them least, while still providing adequate safeguards for employees and members of the public.

Guidance on a wide range of subjects is published by the HSE, the main purpose of which is to help people to understand the requirements of the law, to help people comply with the law and to give technical advice. Following such guidance is not compulsory and employers are free to take other action but if they do follow such guidance they will normally be doing enough to comply with the law.

Approved Codes of Practice offer practical examples of good practice and give advice on how to comply with the law by, for example, illustrating the meaning of the phrase 'suitable and sufficient' which is used in certain regulations. In the UK, Approved Codes of Practice have a special legal status in that if employers are prosecuted for a breach of health and safety law, and it is proved that they have not followed the relevant provisions of the Approved Code of Practice, a court can find them at fault unless they can demonstrate that they have complied with the law in some other way.

Regulations in the UK are law, approved by Parliament, and are usually made under the Health and Safety at Work Act, following proposals from the Health and Safety Commission. However, some risks are so great, or the control measures so costly, that it would be inappropriate to leave employers discretion in deciding what to do about them. Regulations are used to identify these risks and set out specific action that must be taken and, often, these requirements are absolute, that is, something to be done without the qualification of as far as reasonably practicable.

The UK experience over almost 30 years would suggest that the goal setting approach to the management of work place health and safety risks can be very effective, provided that appropriate guidance is provided as to the interpretation of the goals to be achieved and to what constitutes best practice in any given area.

(Contributed by Dr. Joe Gray, WrightWay Training)

SETTING SAFETY TARGETS

Introduction

The implementation of a safety management system is directed to containing or improving safety. For the system to have meaning aims need to be specified and communicated, otherwise there is no sense of purpose in having safety management.

This annex examines a number of objective indicators which can be used to give structure and purpose to a safety management system. They are not exhaustive and companies with different styles of management will need to adopt approaches appropriate to their size and working methods.

Example 1

Captain S. Noonan of Tarquin Shipping, in the last chapter, comments that the commercial benefit of introducing a safety management system is that it has enabled his company to reduce significantly its damage and claims. This then is a reasonable place to start by examining all the damages sustained over the past five years and reviewing what steps can be taken to prevent accidents in future.

Example 2

Captain Hans-Herman Diestel of Alpha Ship, a small to medium sized container fleet, observes that the main cause of loss or damage occurs in pilotage waters. His solution is to improve passage planning and requires his senior officers to attend Bridge Team Management Training. He comments, *"In a small company we have to concentrate on what we believe to be important. We do not really need a sophisticated database of lost time accidents, as we are not really big enough to compile meaningful statistics. Instead, we examine significant injuries which are very rare and I send a fleet memo with the details of all incidents affecting the safety of our operations."*

(ref: SEAWAYS, The Journal of The Nautical Institute, October 2003)

Example 3

Captain Graham Robson, Safety Manager Chevron Tankers, writing in the Institute's book *The Management of Safety in Shipping*, stated that safety management improvement has to be considered in phases. Phase 1 accidents are prevented with improved technology, construction procedures and quality control. Phase 2 accidents occur because of equipment failure caused by misplaced company and/or vessel priorities. Phase 3 are evolutionary, caused by equipment misuse and reduced primarily by on board training. Phase 4 are accidents due to unsafe procedures and Phase 5 are inaction to danger responses by individuals. Phase 6 are categorised as careless mistakes.

Chevron's safety programme is designed to develop Phase 6 safety awareness, in order to eliminate all unsafe actions and conditions. In 1990, when the article was written, lost time accidents per 200,000 man hours had been reduced from 1·8 in 1972 to 0·81 in 1989. To put these figures in context, a typical worker will work 100,000 hours in a lifetime.

(ref: The Management of Safety in Shipping, The Nautical Institute, 1992)

Example 4

Du Pont is the explosives company attributed as the originator of the theory that for every fatality there are 30 near misses, 300 hazardous occurrences and 3,000 contributory minor unsafe acts. It is argued that if these unsafe acts can be captured and recorded, as Captain Stuart Nicholls describes in his chapter, then the process will act to alert workers and crew, lost time injuries will be minimised and major accidents will be avoided.

Du Pont's total recordable injuries per 200,000 hours worked average over the past five years are as follows: Du Pont staff 0·24, off-the-job staff 0·42 and contractors on Du Pont projects 0·61. In the Du Pont magazine No 2, 2003, the following target statement is made:

We are committed to:—

- The highest standards of performance and business excellence.
- The goal of zero injuries, illnesses and accidents.
- The goal of zero waste and emissions.
- The conservation of natural resources, energy and biodiversity.
- Continuously improving our processes, practices and products.
- Open and public discussion, influence on public policy.
- Full management and employee commitment and accountability.

Example 5

There are a number of administrations that keep records of fatalities but few which record them in terms of quantifiable norms. The

Singapore fleet has maintained records which should be considered in the context that one major accident can skew the mortality rate for any particular year. In spite of this, with some 20,500 seafarers employed, the mortality rate per 10,000 man years is about 2·9. The aim of good safety management is, of course, to have no fatalities, but where is the company to start from?

Annualised mortality rates per 10,000 man years

for seafarers in the Singapore fleet (1986—95)

Year	No. of seafarers	Mortality rate all causes	Mortality rate casualty	Mortality rate occupational accidents
1986	9,786	80·73	63·36	4·09
1987	9,771	35·82	20·47	7·16
1988	10,618	58·39	35·79	7·53
1989	11,197	27·69	13·40	4·47
1990	12,710	13·15	0·00	2·47
1991	14,229	31·63	26·79	3·51
1992	15,695	19·11	7·65	1·91
1993	17,667	14·15	5·09	3·96
1994	18,964	15·82	2·64	2·64
1995	20,534	9·74	0·49	2·92

(ref: Doctoral thesis, D. Nielsen, University of Wales, Cardiff, January 2000)

Example 6

Captain Trevor Bailey, shipmaster, observes that all seafarers conduct their work to avoid accidents, but may be unaware of the risks they are taking. The quality and safety manager of a large ship management company asked sea staff to assess risks at sea and the results varied widely and were not generally useable. There is documentary evidence from safety research that risk assessment can be subjective, with high risk takers underestimating risk and low risk takers overestimating. Examining tasks and considering procedures which are safe is a practical way of avoiding risk.

Risk management

Risk management is about planning and considering contingencies and monitoring progress. It is accompanied with exercises and drills and a process for bringing forward matters of concern at ship safety meetings. Where a new activity is being considered it can be helpful to work through a formal process:

- Identify risks and hazards.
- Prioritise management controls.
- Ensure that performance is within the margins of safety.

Attributes of risk control measures are included in a paper by Mr. Jim Peachey, as follows. They can be helpful when trying to articulate the focus for discussion and communication.

1 Category A attributes
1.1 Preventive risk control is where the risk control measure reduces the probability of the event.
1.2 Mitigating risk control is where the risk control measure reduces the severity of the outcome of the event or subsequent events, should they occur.

2 Category B attributes
2.1 Engineering risk control involves including safety features (either built in or added on) within a design. Such safety features are safety critical when the absence of the safety feature would result in an unacceptable level of risk.
2.2 Inherent risk control is where at the highest conceptual level in the design process, choices are made that restrict the level of potential risk.
2.3 Procedural risk control is where the operators are relied upon to control the risk by behaving in accordance with defined procedures.

3 Category C attributes
3.1 Diverse risk control is where the control is distributed in different ways across aspects of the system, whereas concentrated risk control is where risk control is similar across aspects of the system.
3.2 Redundant risk control is where the risk control is robust to failure of risk control, whereas single risk control is vulnerable to failure of risk control
3.3 Passive risk control is where there is no action required to deliver the risk control measure, whereas active risk control is where risk control is provided by the action of safety equipment or operators.
3.4 Independent risk control is where the risk control measure has no influence on other elements.

3.5   Dependent risk control is where one risk control measure can influence another element of the risk contribution tree.

3.6   Involved human factors is where human action is required to control the risk but where failure of the human action will not in itself cause an accident or allow an accident sequence to progress. Critical human factors is where human action is vital to control the risk either where failure of the human action will directly cause an accident or will allow an accident sequence to progress.

3.7   Where a critical human factor is assigned, the human action (or critical task) should be clearly defined in the risk control measure.

3.8   Auditable or not auditable reflects whether the risk control measure can be audited or not.

3.9   Quantitative or qualitative reflects whether the risk control measure has been based on a quantitative or qualitative assessment of risk.

3.10   Established or novel reflects whether the risk control measure is an extension to existing marine technology or operations, whereas novel is where the measure is new. Different grades are possible, for example the measure may be novel to shipping but established in other industries or it is novel to both shipping and other industries.

3.11   Developed or non-developed reflects whether the technology underlying the risk control measure is developed both in its technical effectiveness and its basic cost. Non-developed is either where the is not developed but it can reasonably be expected to develop, or its basic cost can be expected to reduce in a given time scale. The purpose of considering this attribute is to attempt to anticipate development and produce forward looking measures and options.

(ref: Managing Risk in Shipping, The Nautical Institute, 1999)

The information in this annex has been compiled by Mr. C.J. Parker, Secretary, The Nautical Institute 1973—2003.

ANNEX 3

www.ismcode.net

The ISM website was originally set up as an integral part of the research project launched in April 2001 and was sponsored by the North East branch of The Nautical Institute. The site proved to be invaluable as a data collecting facility — where respondents could complete their questionnaires on line as well as air their views about ISM and generally share information. It also became a focus for obtaining information about developments related to ISM implementation.

Unfortunately, because of the intense workload analysing data, preparing the doctoral thesis, writing the book (which was to become *Cracking the Code*) as well as maintaining a full time job, a part time job and trying to be a husband and dad, I am afraid that I did not maintain the site as up-to-date as I would have wished. Some preliminary findings were posted in 2002, comments continued to be received on the discussion forum page and the site continues to attract hundreds of hits every week.

Now that the Doctorate has been completed and the book published I will now have time to devote to not only updating the website but also to completely redesigning and upgrading it to provide a genuine international focus for ISM debate.

This book — *Cracking the Code* — will, hopefully, act as a challenge and a catalyst to encourage many people within our industry to rethink their attitude and understanding of the ISM Code. It will acknowledge that there have been, and still are, many problems with ISM implementation. It will explain how and why many of those attitudes and misunderstandings came about. Much more importantly though it will declare that there is not only a way of overcoming those attitudes and misunderstandings but also that there is a way of developing and implementing a Safety Management System which will make ships safer and, as a natural consequence, make the Company and its ships more efficient and more profitable.

The debate however cannot culminate in this book. For some the light at the end of the tunnel is in sight, for others the tunnel is still very dark and there is a long way to go.

The new website, still with the original web-address — www.ismcode.net — will have, as its main purpose, to encourage others to advance the research into ISM which will help continue to move the debate foreword — to make ships safer and seas cleaner. The site will contain some of the original material, for reference purposes, but will include many new features which will allow the international debate to develop and provide a central reference source for accessing all sorts of information about the ISM Code and its implementation.

The research provided an enormous amount of data and information about the issues surrounding ISM implementation. Inevitably only a small sample of those findings could be brought out in this book. Through the new website, therefore, I hope that much more of that data and information can be made available.

The new website will comprise three main sections:

- The research project.
- ISM Information.
- The ISM Debate.

Those main sections will contain a number of subsections – some brief details of those sub-sections are set out below:

The research project	
The researcher	Some brief details about me and what drives me to encourage the industry to develop a greater understanding of the ISM Code.
Background to the project	Providing an outline, background and details of the research project. Static versions of the original questionnaires will be available for inspection.
Interim report	Some early reflections on the results and findings from the research
Seafarers guide to ISM	Some serious concerns which arose out of the research, involving common misunderstandings about the Code. Some frequently asked questions and some suggested answers/explanations.
Review of the final report	The doctoral thesis *Cracking the Code*. Foreword by the Secretary General of the IMO — Mr. William A. O'Neil

ISM Information	
Primary sources	This will provide a reference to known primary source documents relating to the ISM Code. Any pending or anticipated changes in legislation will also appear here.
Secondary sources	This will include a dynamic reference directory of secondary source documents that will be useful for anyone intending to research any aspect of the ISM Code. Visitors to the website will be invited to submit details of any reference sources of which they are aware and which do not yet feature.
Papers and articles	This section will include a range of research and seminar/conference papers — some published and some that may never have been published which will be made available for general reading and to assist the ISM researcher. Press cuttings will also be included.
Legal reports	Law reports and other legal commentaries on cases that have been the subject of judicial decision and discussion where aspects of ISM have been raised as an issue. Visitors to the site will be invited to submit any reports that might have arisen in their part of the world.
Accident and near miss reporting	Details and links will be provided to accident and near miss reporting sources — such as The Nautical Institute 'MARS' reports, maritime 'CHIRPS', the MAIB reports and similar.
Training and service providers	A dynamic catalogue of suppliers of training in ISM related topics plus services linked to ISM implementation. This will include a link to The Nautical Institute nautical*campus* site. I will not necessarily have vetted the suppliers of the information and I make no assertion at all about any of their products or services. That is for each individual to make their own assessment. Where possible a website or email link will be provided where more information can be obtained.

Links to other sites	Details of other websites which may be relevant and/or interesting to anyone researching ISM.
The ISM Debate	
ISM Chat room	An open forum debating page, where visitors to the site can exchange ideas with myself and with each other. If possible I will answer specific questions, possibly using data from the research database. This may be statistical data or narrative feedback, possibly including information which it was not possible to include either in the Doctoral thesis or in *Cracking the Code*.
ISM Opinion	This page was included on the original website and allowed anyone to express their views about ISM in an anonymous way. The website visitor could add their opinion and read the opinions of others. The sharing of experiences and perceptions in this way allowed others to understand that there were many people experiencing similar problems or successes. This part of the site will continue and visitors will be encouraged not only to share their problems but also to share details of any remedial steps they may have put in place to overcome those problems.
ISM Researchers	There would appear to be many individuals around the world who are actively involved in researching different aspects of the ISM Code or related topics. This page will allow those researchers to post details of their areas of interest and provide an opportunity to meet other researchers who may be involved in similar fields. Details of Universities, Colleges and other Academic institutions where ISM research can be undertaken will also be provided here.

Forthcoming events	Details of ISM related seminars, conferences, training courses and similar can be made available on this part of the website with appropriate links to the organisers.
The lighter side of ISM	The ISM Code and safety generally are very serious subjects. However, even such a serious subject can have its lighter side. This part of the site will allow us to reflect on the funny side of ISM. Website visitors will be invited to submit their ISM jokes, funny stories, anecdotes, poems, cartoons or pictures to share with others.

The success or failure of the ISM Code lies in the hands of every single individual in the industry. The ship operator needs to develop inspirational leadership and make the necessary commitment in terms of money, people and time. The seafarers and the staff ashore need to make their own commitment to the implementation process, to work together as one team and develop a Safety Management System that will make ships safer, more efficient and more profitable. Flag State and Port State Authorities, along with the Classification Societies and other Recognised Organisations need to work constructively with the industry and help them develop their SMSs and reward those who are genuinely trying. Similarly, Judges, Arbitrators, lawyers and insurers should also encourage and help those who are genuinely trying and make it clear that anything less will not be tolerated.

There are many hurdles still to overcome. I hope that my website will, in some way, contribute towards these goals and help to keep the debate moving foreword in a healthy and positive way. ISM, by its very nature is a dynamic force. It will continue to grow and develop, helping to make ships safer and seas cleaner.

The success of the site, however, will depend upon the level of participation by those visiting the site. Please visit the site and become involved — your contribution is important!

The website will continue to be sponsored by the North East Branch of The Nautical Institute.

Phil Anderson
Acomb, Northumberland, August 2003

ANNEX 4

NAUTICAL INSTITUTE SAFETY MANAGEMENT RESOURCES

The Nautical Institute Web Services (www.nautinst.org)

Visit www.nautinst.org for information about the Institute and membership criteria, branch details and links to branch sites, comprehensive listing of all publications and training services, the much acclaimed Marine Accident Reporting Scheme (MARS) public database, relevant conference listings and links to other sites of interest.

Confidential Marine Accident Reporting Scheme (MARS)

The Nautical Institute undertakes to publish each month in its journal a special section on confidential near miss reports from members all over the world. The reports are also freely available on the Institute website (www.nautinst.org). The reports are part of the Institute's contribution to helping the profession to learn from the experience of others. The reports are regularly used for discussion on board in safety meetings and for instruction.

The Nautical Campus (www.nauticalcampus.org)

This is a global career management and learning resource site for all maritime professionals, with four major sections promoting continuous professional development and career monitoring for those in the maritime industry both at sea and ashore.

Courses — being built up to contain the world portfolio of courses offered by educational establishments and training providers. Lists hundreds of courses ranging from STCW standard courses through distance learning ballast management to training for safety management with course providers from all over the world.

Resources —There is a huge and disparate selection of sources of information relating to the marine sector and this part of Nautical Campus links to directories of nautical books, publications, periodicals, publishers and booksellers.

Thinking ahead — This is a section designed to answer the questions which are most frequently asked concerning careers afloat and ashore.

Careers — is an authoritative list of maritime occupations and the qualifications necessary to meet recruitment standards alongside personal accounts of those working in these fields.

Publications

Three relevant books on safety management available from the Institute are: The Management of Safety in Shipping, The Management of Risk in Shipping and Managing Quality and Safety in Shipping. Full details are on the website above.

SEAWAYS

SEAWAYS is the monthly journal of The Nautical Institute, sent out free of charge to all members by mail in the UK and airmail overseas. The journal is the principal organ for keeping the profession up to date and contains features and reports directed at the qualified mariner.

Alert!

An international human element bulletin sent free to all who are interested in this subject. It aims to link all the professional maritime disciplines and is supported by a website (www.nautinst.org/alert) which enables contributors to enter papers and contacts to advance the understanding of human element issues.

INDEX

BIOGRAPHIES

Dr. Phil Anderson DProf BA(Hons) FNI Master Mariner

Dr. Anderson is widely recognised internationally as a leading authority on the ISM Code and safety management on board ships. This volume is largely a product of a major piece of research he undertook during 2001 on ISM implementation between the Phase One and Phase Two implementation deadlines. That research also formed the basis for his doctoral submission to the National Centre for Work Based Learning Partnerships at Middlesex University.

A Master Mariner, Phil served for near 12 years at sea with the Bibby Line of Liverpool. Coming ashore in 1980 he has spent the last 23 years working within P&I Clubs and is presently a full director of the Management Company of North of England P&I Association in Newcastle upon Tyne. Within the industry he is recognised for his pioneering work in developing loss prevention as an integral part of the service provided by P&I Clubs.

Currently senior Vice President of the Nautical Institute, Dr. Anderson is due to assume the Presidency in May 2004. An avid supporter of the aims and principles of the Institute for many years Phil devotes much of his time to local, national and international Nautical Institute activities. In 1989 he led a team which produced the most successful book ever published by the Institute: *The Master's Role in Collecting Evidence*, subsequently retitled *The Mariner's Role in Collecting Evidence*. He also co-authored the Institute publication *Accident and Loss Prevention at Sea* in 1993.

Following an approach from Lloyds of London Press, Phil wrote a volume for their practical guide series — *ISM Code–A Practical Guide to the Legal and Insurance Implications* — which was published in 1998.

In 2002, as a consequence of results coming back from the research, and in collaboration with Peter Kidman, he produced two further ISM related volumes: *A Seafarers Guide to ISM* and *What have the World Cup and ISM got in common?* – both volumes published by North of England P&I Association. When the opportunity allows, Phil can be found on board his 38 foot wooden sailing cutter, Ida Maria, off the North East coast.

Captain Stuart Nicholls MNI

Since leaving school for a life at sea in 1988 Captain Stuart Nicholls has achieved much in his fifteen years of seagoing experience. From cadet to Master in eleven years, Captain Nicholls assumed his first command at the age of twenty seven.

Captain Nicholls was involved in the introduction and implementation of the ISM code at sea. Since transferring his skills to the offshore sector in 2001 he has been actively involved in the real time risk management of incidents and occurrences through proactive reporting and the implementation of corrective action procedures offshore.

The experience gained whilst deep sea and latterly in the offshore sector has enabled him to compare and contrast the very best and worst that safety management has to offer. By contributing to this book he hopes to impart to the reader, some of the lessons and observations he has experienced to date, in an easy to understand language that will hopefully communicate his message to all rank and nationality of officer and crew.

This year Captain Nicholls will assume the role of managing director of a new and exciting company, providing experience based training, focused on the need to understand what the safety and commercial benefits are for those who dare to embrace a company wide policy of, "total safety".

Captain John Wright FNI MIOSH AMIQA

John is the Managing Director of WrightWay Training Limited (WWTL). As a Master Mariner he has wide, practical management experience in the marine and oil industries and over the last eleven years has focused particularly on Safety Management. WWTL is the UK representative of the Danish Maritime Institute and John is responsible for presenting their accredited Crew Resource Management programme. Clients include the marine and offshore oil industries and also shore based organisations. His specialist areas are human factors, safety and business culture change, risk assessment and accident investigation.

These skills are used to support organisations in their quest for safe, accident free working environments, which in turn leads to business excellence.

Captain Sean Noonan FNI

Captain Sean Noonan FNI is a Marine Consultant and a certified Auditor in Safety and in Quality, working on a self employed contract basis with his company Faire Marine Services Ltd. Much of his work involves conducting risk assessments on tankers with Total's ship inspections department and for the OCIMF. He also provides consultancy advice to ship operators who may be experiencing difficulties getting the best out of their Safety Management Systems.

Captain Noonan was previously the Designated Person Ashore and Marine Superintendent in charge of the Superintendency Department of Tarquin Gas Carriers Ltd. (TGC), a Scottish ship owning company. There he oversaw the setting up and running of their Safety Management System (SMS). During the writing, implementation and running of the SMS for TGC, he also oversaw the Company's auditing and additional training programme for ship staff and shore staff and conducted shore based seminars.

Prior to coming ashore in the year 2000 Sean had followed a seagoing career for twenty years, latterly sailing with Worldwide Shipping as Master. Captain Noonan shares some of his valuable and wide-ranging experience in this book — *Cracking the Code*. Speaking from first hand experience he demonstrates that, properly implemented, ISM can be used to make operating ships not only safer and more efficient but more profitable.